A
DICTIONARY
OF
MATHEMATICS

A
DICTIONARY
OF
MATHEMATICS

ROY HOLLANDS

*Department of Mathematics,
Dundee College of Education*

Longman

LONGMAN GROUP LIMITED
London
*Associated companies, branches and representatives
throughout the world*

First published 1980
ISBN 0 582 18080 5

**British Library Cataloguing
in Publication Data**

Hollands, Roy Derrick
   A dictionary of mathematics.
   1. Mathematics – Dictionaries
   I. Title
   510′.3   QA5   78-41312

ISBN 0-582-18080-5

Set in 9/11 pt Times, Monophoto 327

Printed in Hong Kong
by Sheck Wah Tong Printing Press Ltd

ACKNOWLEDGEMENTS

The publishers are grateful to the following for permission to reproduce photographs: Barnaby's Picture Library, pages 1 (photo: S. B. Davie), 5 (photo: E. Preston), 22 right, 23, 40, 99 (photo: R. F. Edwards), 109 below (photo: Peter Raistrick), 134 left (photo: Mike Jay), 134 right (photo: T. H. Williams), 139 (photo: W. R. Bowan), 140 (photo: B. Alfieri) and 143; British Museum, page 38; British Railways, page 130 right; Camera Press, pages 115 (photo: Ralph Crane) and 123 (photo: Ken Lambert); ESA Creative Learning Ltd, pages 10, 11, 26 above, 47, 49, 98, 109 above, 112, 146 and 147; Keystone Press Agency, page 81; London Fire Brigade, pages 50 right and 132; Mansell Collection, pages 7, 50 left, 54, 63, 101, 117 left and 124 and NASA, page 92. The map on page 91 is reproduced from the Ordnance Survey 1:50,000 map with the permission of The Controller of H.M.S.O. Crown © Reserved.

A NOTE ON THIS BOOK

Many new words have appeared in school mathematics during the last twenty years, particularly those associated with modern maths.

A Dictionary of Mathematics uses clear and simple language to define with mathematical accuracy both modern and traditional terms. There are over 600 entries. These cover both familiar and more complex mathematical terms and also short accounts on mathematicians such as Piaget and Einstein.

To enable the reader to reach a deeper knowledge of the terms, there are numerous photographs and illustrations, examples of usage and other relevant interesting information. The origins of some words are given and useful formulae, symbols and tables are included. The book is designed to be within the mathematical interest and understanding of children aged 10+. The words printed in bold are entries that can be found in the appropriate place in the Dictionary.

The comprehensive approach makes *A Dictionary of Mathematics* a useful and attractive source of reference for children, teachers and parents in school and at home.

ABACUS

A calculating device consisting of beads that slide along wires. The beads **represent** 1's, 10's, 100's, etc according to which wire they are on. The abacus is particularly helpful as an aid to learning **place value**.

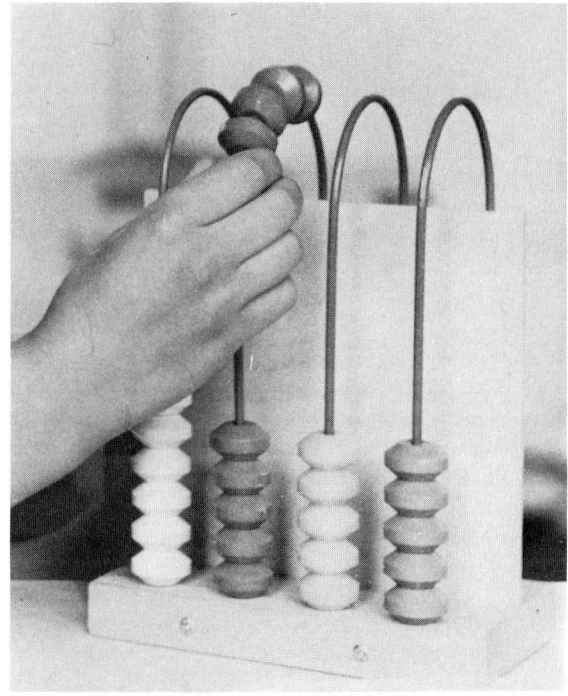

In ancient forms pebbles were placed in grooves made in the sand. Later sand trays enabled people to carry the abacus around. There are many different forms, including the **Roman abacus, Russian abacus, soroban** (Japanese) and **suan-pan** (Chinese).

The picture shows a girl using a hoop abacus.
LATIN *abacus*, a board for calculating on.

ABBREVIATION

A shortened form.
cm is the abbreviation for centimetre,
m² for **square metre**, 5³ for 5 × 5 × 5.

ABSCISSA

The shortest distance from the **vertical axis (y-axis)**.

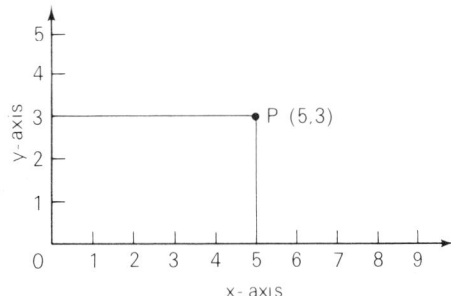

In the diagram the abscissa of P is 5. The distance of P from the **x-axis** is called the **ordinate**. P's ordinate is 3. 5 and 3 are called the **coordinates** of P. They are written as an **ordered pair** (5, 3). The abscissa is the first of these two numbers, the ordinate is the second.

ACCELERATION

The **rate of** change of velocity with respect to **time**. (See VELOCITY.)
Example: A car going along a **straight** road changes its velocity from 80 cm per **second** to 100 cm per second in 4 seconds. The change in velocity is 20 cm per second in 4 seconds, so the **average** acceleration is 5 cm per second per second. This is written as $5 \, cm/s^{-2}$ (or $5 \, cm/s^2$).

ACCOUNT

A statement of money or goods that are owing or due.

ACCURATE

1 Exactly right. Without **error**.

2 No measuring instrument can be exactly correct. We therefore **measure** to a certain degree of accuracy.
The **length** of a pen might be measured to the nearest **millimetre** (0.1 cm). If given as 8.3 cm this tells us that the pen's length is nearer to 8.3 cm than it is to 8.2 cm or 8.4 cm. It is accurate to the nearest millimetre.

ACRE
A measurement of area.

1 acre = 4840 **square yards** or $\frac{1}{640}$ square **mile**.

In **metric** units an acre is **approximately** 4047 square **metres** or 40.47 ares.

The Roman word for field was 'ager' and from that we get acre. It was the amount of land that a pair of oxen could plough in one day. Land used to be cultivated in strips of length one **furlong** (furrow long). At the end of each strip (220 yards) the oxen needed a rest before turning. The width of each acre was 22 yards.

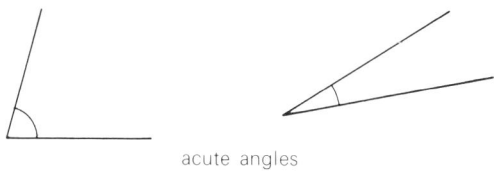

At first an acre was always the same **rectangle**, 220 yards by 22 yards. Later only the amount of ground was considered and it could be any shape at all. Forest was not as valuable as good farmland so in many places an acre of forest was larger than an acre on which crops could be grown.

ACUTE ANGLE
An **angle greater than** 0° but **less than** 90°.

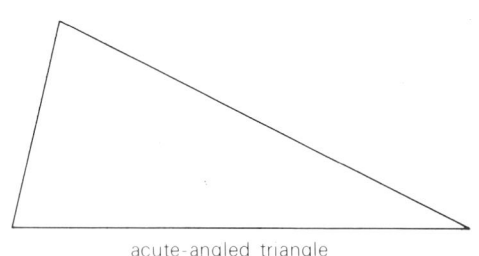

acute angles

LATIN *acutus*, sharpened.

ACUTE-ANGLED TRIANGLE
A **triangle** with all three **angles acute**. It is sometimes called an acute triangle.

acute-angled triangle

AD
Abbreviation for 'anno Domini' which means 'in the year of our Lord'. AD shows that the number of years is after the birth of Christ. The Battle of Hastings took place in 1066 AD. (See BC.)

LATIN *ano*, in the year; *dominus*, master.

ADD
See ADDITION.

ADDEND
Any one of a **set** of **numbers**, or **quantities**, that are to be added.

$13 + 9 = 22$ 13 and 9 are addends.

$5\,cm + 8\,cm = 13\,cm$ 5 cm and 8 cm are addends.

ADDITION
The operation of combining several quantities to form a further quantity, called the **sum**.

If only two quantities are combined it is called a **binary operation** as in $3 + 4$.

To add $5 + 2 + 6$ we have a sequence of binary operations, $5 + 2 = 7$ and $7 + 6 = 13$.

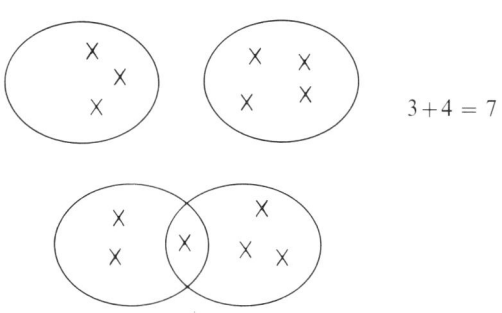

$3 + 4 = 7$

Although the **sets** have 3 and 4 x's as before there are only 6 altogether.

When adding **numbers** the sets they refer to must not overlap.

(See DISJOINT SETS.)

ADDITIVE INVERSE See INVERSE.

ADDRESS
In a **computer**:

1 The label that gives the position of stored material. This could be the number of the register holding the material or the position of a magnetic core that is storing the facts.

2 The position of a **point** in a **plane** given in terms of its **coordinates**.

The address of P is $(4, 2)$.

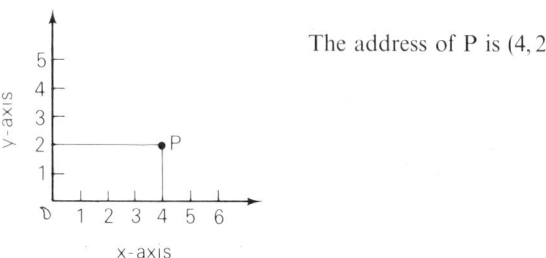

ADJACENT ANGLES

Adjacent means 'lying next to'.
Adjacent angles are those with a **common vertex** and a common **line**.

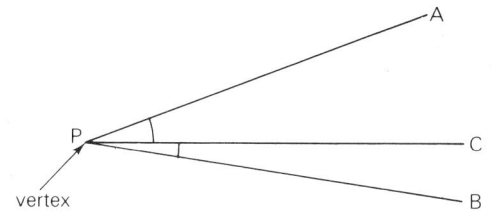

P is the common vertex. PC is the line common to angle APC and angle BPC.

AGGREGATE

1 A whole **collection** as opposed to the separate parts.
Example: All the **even numbers**.

2 The **total** or **sum** when numbers are involved.
Example: A football team scores 32 goals in a season. It has an aggregate of 32 goals.

ALGEBRA

Elementary algebra is the study of **number** systems and their **properties** in a general way. Letters or **symbols** are used to **represent quantities**, and **signs** to represent the connections between them.
Algebra extends **arithmetic**. For example when any two numbers are added this can be represented by $a+b$ instead of all the particular cases such as $3+4, 2+8$, etc.
A special branch of algebra, the algebra of **sets**, uses expressions such as $A \cup B$ (A **union** B), $A \cap B$ (A **intersection** B).
More advanced algebra includes **calculus**, **logic**, number **theory** and many other topics. Symbols were used to represent numbers from about 1700 BC but the word algebra probably comes from the Arabic work entitled Ilm al-jebr-wal muquabalah (AD 825). This means 'putting together broken parts'.

ALGORITHM

Also called algorism.
A systematic procedure for finding the **solution** of a **problem**. Each step is clearly laid out.
Example: Algorithm for multiplying a two-**digit number** by 7. (The working for 48×7 is given to make the method clear.)
a **Multiply** the units digit by 7 ($8 \times 7 = 56$)
b Write down the units digit
  from your answer in *a* (6) 48
  $$\begin{array}{r} 48 \\ \times\ 7 \\ \hline 6 \end{array}$$

c Multiply the tens digit of the two-digit number by 7. ($4 \times 7 = 28$)
d Add the tens digit from your answer in *a* to the answer from *c* ($28+5 = 33$)
e Write down this on the left of
  b (336) 48
  $$\begin{array}{r} 48 \\ \times\ 7 \\ \hline 336 \end{array}$$

An algorithm is not needed for simple **calculations** but can be a great help in complicated work.
ARABIC Derived from *al-khowarizmi*, an Arab mathematician of the 9th century.

ALIQUOT PART

Any part of a **quantity** that is an exact **divisor** of the whole.
20 **centimetres** is an aliquot part of 1 **metre**. 4 is an aliquot part of 16.
An aliquot part is always a **unit fraction** of the whole, that is a fraction with 1 in the numerator.

20 cm is $\frac{1}{5}$ of 1 m.   4 is $\frac{1}{4}$ of 16.

ALTERNATE ANGLES

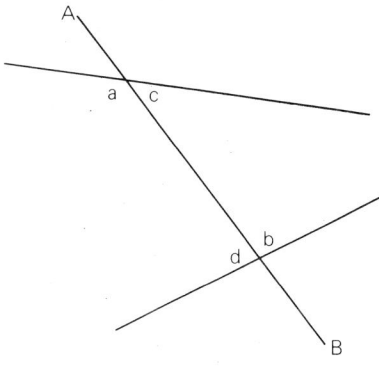

A **straight line**, AB, cuts two other straight lines.
The **angles** a and b are called alternate angles. c and d are also alternate angles. When two **parallel** lines are cut the alternate angles are equal.
$e = f$ and $g = h$.

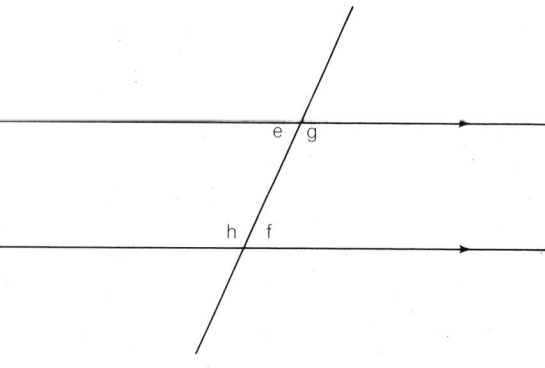

ALTITUDE

1 The **height** of a place above sea-level.

2 For a **plane figure**.
A **line** from a **vertex perpendicular** to the opposite **side**.

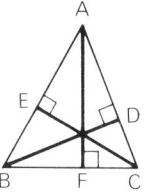 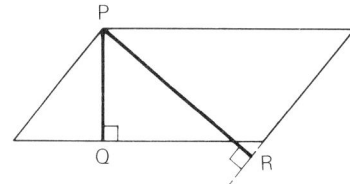

A **triangle** has three altitudes.
AF, BD and CE are altitudes.

From P two altitudes can be drawn, PQ and PR.

3 For a **solid**.
A line from a vertex perpendicular to the opposite **face**.

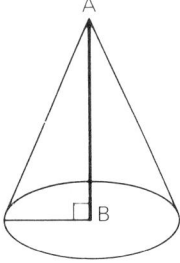

AB is the altitude of this **cone**.

4 The term altitude may be applied to the **length** of the lines instead of the lines themselves.
Example: The altitude of this triangle is 5 cm.

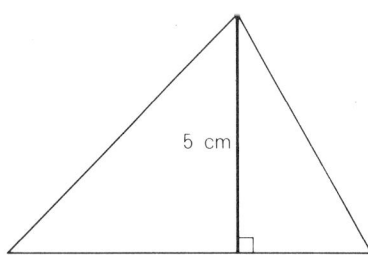

5 Angle of altitude.

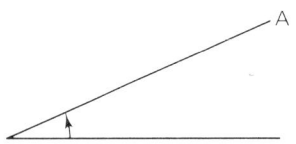

The angle between the **horizontal** and a line from the observer to the object (A).
Also called ANGLE OF ELEVATION.

A.M. or a.m.

Abbreviation for Ante Meridiem, or Antemeridian, before **mid**day. Any time between midnight and the following noon (12 a.m.) on the same day.
LATIN *ante*, before; *meridiem*, noon.

AMBIGUOUS CASE

When two **sides** and one **angle** of a **triangle** are given, four possible triangles may be drawn according to the **lengths** of the sides, the **size** of the angle and the relative **positions** of the sides and the angle. The ambiguous case occurs when the side opposite the given angle can be drawn so as to give two different triangles. For example,
angle A = 30°, AB = 6 cm, BC = 3 cm.
Triangles BAC_1 and BAC_2 both satisfy the given conditions.

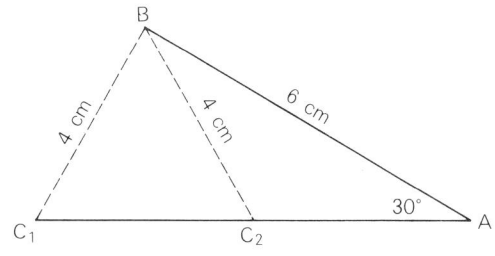

AMICABLE NUMBERS

Two **numbers** such that (i) the **factors** of the first (except for the number itself) add up to give the second number, and (ii) the factors of the second (except the number itself) add up to give the first number.
Example: The factors of 220 (not including 220 itself): 1, 2, 4, 5, 10, 11, 20, 22, 44, 55 and 110. The **sum** of these is 284. The factors of 284 (not including 284 itself): 1, 2, 4, 71 and 142. The sum of these is 220.
284 and 220 are therefore amicable numbers.
Pythagoras (582–507 BC) is said to have discovered this property of 220 and 284 and it was not until 1636 that a second pair was found, 17 296 and 18 416.
Euler in 1750 added fifty-nine more pairs. A sixteen year old boy was the first to find that 1184 and 1210 were amicable.
Amicable numbers (also called amiable) were thought by the Arabs to bring friendship to people possessing them.

AMOUNT

1 A general term for **quantity**. As in the amount of sand needed to fill a bucket.

2 The **sum** of two or more quantities, especially money.
Example: A man owes £48 and £16. The amount of his debt is £64.

3 When referring to a loan or investment.

principal (money borrowed) + **interest** = **amount**.

Example: £100 principal is lent at 5 **per cent simple interest** for 2 years.

Interest = £10.
Amount = principal + interest = £100 + £10 = £110.

AMPERE

A **unit** for measuring electric current in the S.I. system. Named after a French mathematician and physicist of that name (1775–1836).

ANALOGUE COMPUTER

A **machine** for **calculating** in which **numbers** are represented on some form of **scale** and not by counting things. A **slide rule** is an analogue computer where **length** represents numbers.

Thermometers and **speedometers** are other examples. (See DIGITAL COMPUTER.)

ANALYSIS

1 The process of separating into parts. These parts are often traced back so as to see what they depend upon.

2 A branch of advanced **mathematics**. It deals with the **infinitely** large and infinitely small and includes **calculus**, **functions**, **limits**, **series** and **convergent sequences**.

ANGLE

The **measure** of an **amount** of turning or **rotation**.
Strictly: The **figure** formed by two **lines** that have a common **point** or **vertex**. The lines (or **rays**) are called the sides (or **arms**) of the **angle**.

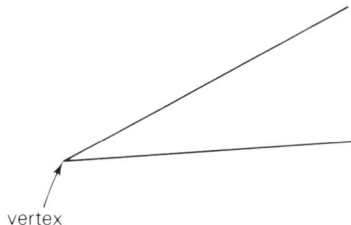

vertex

(See ACUTE ANGLE, ADJACENT ANGLE, ALTERNATE ANGLE, COMPLEMENTARY ANGLE. CORRESPONDING ANGLE, OBTUSE ANGLE, OPPOSITE ANGLE, REFLEX ANGLE, RIGHT ANGLE, STRAIGHT ANGLE, SUPPLEMENTARY ANGLE.
See DEPRESSION and ELEVATION for ANGLE OF DEPRESSION and ANGLE OF ELEVATION.)

ANGLE OF DEPRESSION

See DEPRESSION.

ANGLE OF ELEVATION

See ELEVATION.

ANNULUS

The **region** between two **circles** with the same **centre** (**concentric** circles).

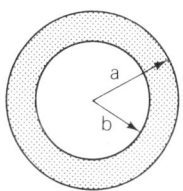

If the **radii** of the circles are a and b (a > b) the **area** of the annulus is $\pi a^2 - \pi b^2$ or $\pi(a^2 - b^2)$.

ANTE MERIDIEM or ANTEMERIDIAN

See AM.

ANTI-CLOCKWISE

A **rotation** in the opposite **direction** to the movement of a **clock's hands**. Also called **counter-clockwise**.

GREEK *anti*, against.
LATIN *cloca*, a bell. A bell was rung to denote the **time** of **day**.

APEX

The **point** of a **figure** that is the greatest distance from its **base**.

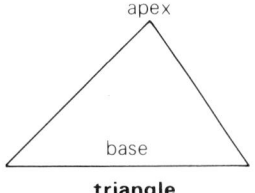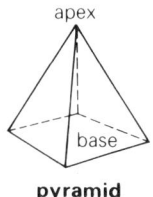

triangle   **pyramid**

LATIN *apex*, top or point.

APPLIED MATHEMATICS

Mathematics that can be used in a practical way as in physics, surveying or engineering. It is also used for mathematics that is based on collecting information or measuring, then **theories** are developed so as to take the ideas to a more advanced level.

Applied mathematics is particularly concerned with **mass**, energy, **force**, **time** and **space**. In the past mathematics has been separated into either **pure mathematics** or **applied mathematics**.

APPROXIMATION

A **number** which is accepted as near enough to another number for the purpose. For example, π(pi) may be taken as $\frac{22}{7}$ or 3.14 whereas it is 3.141592... with the **digits** never repeating and never coming to an end. No measurement is ever **exact** so all measurements are approximations.

The **sign** for approximation is $\simeq$, $\approx$ or $\doteqdot$. We would write $\pi \approx 3.14$; $6.892135 \simeq 6.9$.

ARABIC NUMERALS

The **numerals** invented by the Hindus were modified by the Arabs and then introduced to Europe about 700 years ago. A European manuscript written in the 13th **Century** shows the following numerals.

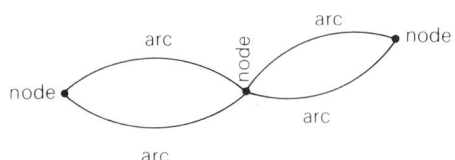

(See HINDU-ARABIC SYSTEM.)

ARBITRARY UNITS

Units of **measurement** that are not in general use throughout a country. They may all be the same **size** such as orange **Cuisenaire** rods, marbles (for **weighing**) or postcards (for **area**). They may vary from one user to another as with body measurements (**span**, **cubit**, **pace**).

ARC

1 Part of a **curve** but not the whole curve. Arc can be applied to any curve but is most frequently met with reference to a **circle**. A **semicircle** is an arc which is **half** of a circle.

A **minor** arc. An arc **less than** a semicircle.

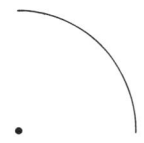

A **major** arc. An arc **greater than** a semicircle.

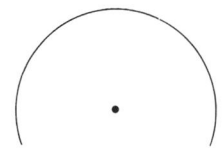

An arc could be defined as a **subset** of the **set** of **points** on a curve.

2 In **topology**, arcs are the **lines** in a **network**. They join the points (or **nodes**).

LATIN *arcus*, a bow.

ARCHIMEDES, 287–212 BC

Archimedes was the most famous mathematician and scientist of ancient times. He was born in Syracuse in Sicily, his father being an astronomer. His main work was on **areas**, **volumes** and **centres** of **gravity** and he began the branches of **mathematics** now known as **calculus**, hydrostatics and **mechanics**. His work in mechanics included a study of the **lever** and **pulleys**. Archimedes put his ideas to practical use by inventing war machines that the Romans greatly feared.

The picture shows the death of Archimedes when the Romans captured Syracuse. He was working out a **problem** in the sand when a soldier demanded that he should go to the Roman commander, Marcellus. Archimedes said he had not finished his problem and continued with it whereupon the soldier drew his sword and killed Archimedes.

ARE

A unit of **area** in the **metric system.**
1 are = 100 **square metres** (100 m²).
1 are is approximately 119.6 square **yards**.
LATIN *area*, an open **space**.

AREA

The **amount** or **size** of a **surface**.
The **measure** of a closed **region**. This is measured in square **units** such as **square centimetres** (cm²) or square **metres** (m²).

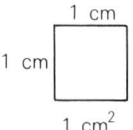 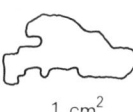

LATIN *area*, an open **space**.

ARITHMETIC

The study of **number**.
Generally used for **computations** with **whole numbers** and **fractions**, including **decimals**. The most common computations are **addition**, **subtraction**, **multiplication** and **division** but there are others, such as finding the **square root**.
Arithmetic is also used for all **numerical** aspects of **mathematics**. **Measurement**, solving number **problems** and paying for goods are therefore included.

ARITHMETIC MEAN

The result of adding **numbers** or **quantities** and dividing the **sum** by the number of **terms**.
Example: 6, 3, 8, 7.

$$\text{Arithmetic mean} = \frac{6+3+8+7}{4} = \frac{24}{4} = 6.$$

Also called arithmetic **average** but should not be called just average as this applies also to **mode**, **mean** and **geometric mean**.
The term is also used for the missing term in an **arithmetic progression**.
Example: 2, 5, 8, 11, −, 17. The missing term is the arithmetic mean of 11 and 17.

$$\frac{11+17}{2} = \frac{28}{2} = 14.$$

ARITHMETIC PROGRESSION

A **sequence** of **numbers** in which the same number is added (or subtracted) to give the next **term**.
Example:
a 7, 12, 17, 22, 27, 32, 37, . . . (5 is added).
b 78, 72, 66, 54, . . . (6 is subtracted).
The **amount** added is called the **common difference**
(*a*) has a common difference of 5.
(*b*) has a common difference of −6.

(Note that subtracting 6 can be thought of as adding −6). The **sum** (S) of a progression is given by the **formula**

$$S = \frac{n}{2}[2a + (n-1)d].$$

n = number of terms
a = 1st term
d = common difference.

Example: Find the sum of the numbers 1 to 100.

n = 100, a = 1, d = 1.
S = $\frac{100}{2}[2 \times 1 + (100-1)1]$
= 50[2 + 99]
= 50 × 101
= 5050.

ARMS OF AN ANGLE

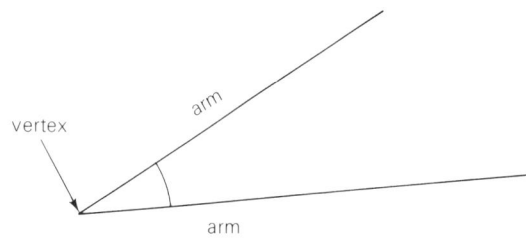

The two **lines** forming an **angle**. They are also called '**sides**'. (Strictly speaking an angle is formed by two **rays**.)

ARRAY

An arrangement in **rows** and **columns**.
These **numbers** form an array.

16	27	43	62	49
28	26	31	16	24
1	43	53	89	73

This is *not* an array

23	43	7	19
	43	12	9
	17	1	

ARROW GRAPH

A **graph** showing **relations** by **arrow lines** joining related **numbers** or **elements** of a **set**.
Example: is **less than**

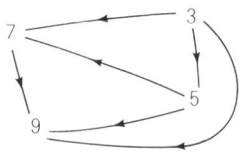

Also called arrowgram.

ARROW LINE

A line with an arrow-head showing **direction**. →— or ⟶.

a John $\xrightarrow{\text{is taller than}}$ Mary. The arrow line shows the **relation** is 'taller than'. The direction is important and John —← Mary is not true for the given relation.
b See the entries for **mapping**, **one-to-one correspondence**, **line** and **ray**. These show other uses of arrow lines.
LATIN *linum*, flax. (Linen is made from flax and a stretched linen thread was used to give a **straight line**.)

ASCENDING ORDER

Increasing: These **numbers** are in ascending order:

2, 8, 9, 26, 30.

Each **term** is **greater than** the one before.

ASSOCIATIVE

The associative law or property:

(a ∗ b) ∗ c = a ∗ (b ∗ c) where a, b and c **represent quantities** and ∗ represents an **operation**, such as **addition, subtraction, multiplication and division**.
a For addition:
   Quantities can be grouped in either of two ways and the **sum** is the same.
Example: 2 + 5 + 3 can be grouped as (2 + 5) + 3 = 7 + 3 = 10 or as 2 + (5 + 3) = 2 + 8 = 10.

b For multiplication:
   Quantities can be grouped in either of two ways and the **product** is the same.
Example: 2 × 5 × 7 can be grouped as (2 × 5) × 7 = 10 × 7 = 70 or as 2 × (5 × 7) = 2 × 35 = 70.
Subtraction is *not* associative. Consider 10 − 3 − 2.

(10 − 3) − 2 = 7 − 2 = 5 but 10 − (3 − 2) = 10 − 1 = 9.

Division is not associative. Consider 16 ÷ 4 ÷ 2

(16 ÷ 4) ÷ 2 = 4 ÷ 2 = 2. 16 ÷ (4 ÷ 2) = 16 ÷ 2 = 8.

ASTROID

A **curve** formed by a **point** on the **circumference** of a **circle** that **rolls**, without slipping round the inside of a fixed circle, making the four-sided **figure** shown below.

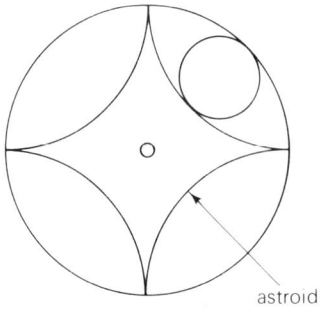

astroid

GREEK *astron*, star.

ASYMMETRY

Not having **symmetry**.

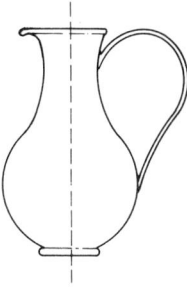

Example: The jug is not symmetrical about the **line** shown as there is a handle on one side but not on the other.
GREEK *a*, not; *syn*, together; *metron*, a **measure**.

ATTRIBUTES

Any **property** for instance colour, **shape**, **size**, **number** or cost.
Example: What attributes have these in common? Pear, orange. They are fruits. They grow on trees. They all have pips. In what attributes do they differ? They are different in shape, colour, thickness of skin and many other ways.
LATIN *ad*, to; *tribuere*, give.

AVERAGE

Another **term** for **mean**. There are several forms of average: **arithmetic mean**, **geometric mean**, **mode**, **median** and others.
Average is often taken, in everyday terms to be the same as arithmetic mean but the wider meaning above is the one a mathematician would use.

AVOIRDUPOIS

A system of **weights** used in the United Kingdom and some other countries before the introduction of **metric units** with the **Système Internationale**. It was used for all goods other than drugs, jewels or precious metals.
OLD FRENCH *aveir de pes*, to have weight.

AXIOM

A **statement** that is assumed to be true and from which other statements are then proved. Axioms are also called 'postulates' or 'assumptions'.
Example: One **straight line** and one only can be drawn through two distinct **points**.
An axiom may seem obvious, or self evident, but there are always some assumptions to be made at the start and these cannot be proved. A proof must follow logically from what is known for certain and not be based on what seems to be true. For instance **parallel** lines *seem* to get closer together as we look at them in the distance. This could mislead us into thinking that they do actually meet.

AXIS

1 A fixed **line** such that the position of any **point** can be **measured** by distances along it. When giving the position of a point in a **plane** we generally use two lines at **right angles** to each other. The **horizontal** one is called the **x-axis** and the vertical one the **y-axis**. Where they meet is called the **origin** (0). The axes are called the axes of **coordinates**.

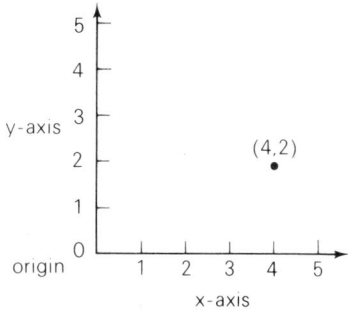

The two **numbers** denoting a point are called coordinates. The first of the pair shows the distance in the **direction** of the *x*-axis (4) and the second the distance in the direction of the *y*-axis (2). They were named **Cartesian coordinates** after René Descartes (1596–1650), a French mathematician.

2 A main line through the **centre** of a **plane** or **solid figure**.

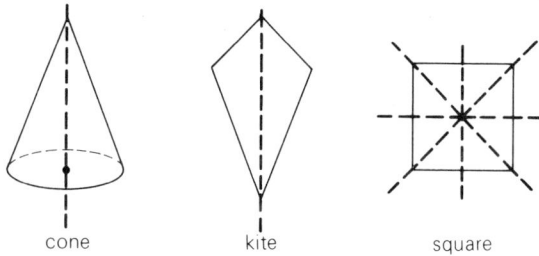

cone   kite   square

These lines may be axes of **symmetry** as above, but this is not always so as shown below:

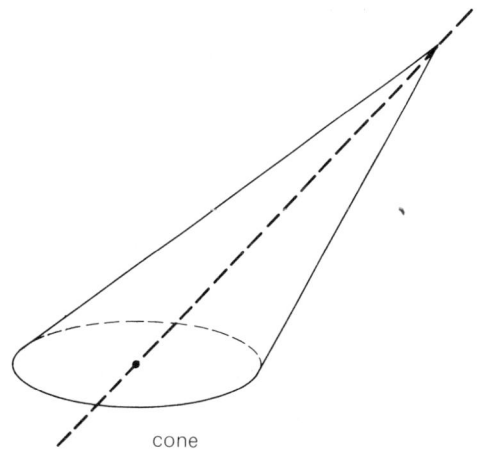

cone

B

BABYLONIAN SYSTEM

A **number** system based on 60. It was used by the Babylonians about 4000 years ago.

As 60 was represented in the same way as 1 some confusion arose but generally you could tell which was meant by common-sense. For example if a man had Y Y sons it was obviously 2. The system was also called the **sexagesimal system**. We still use 60 **seconds** = **1 minute**, 60 minutes = **1 hour**. The 360° in a complete turn also came from the 60 **base** (60 × 6 = 360).

Other connections are 60 original gods in Babylonia and their **division** of their year into 360 days. This error was not particularly important as the seasons do not have marked **differences** as ours do.

BALANCE

1 An **equal distribution**, or **sharing** out.
Example: Good health depends upon having a balanced diet, or
The armies in Europe were evenly balanced.

2 The **amount** still owing or remaining.
Example: He paid £5 for a coat costing £20. This left a balance of £15 still owing, or
After drawing out £20 from the bank he had a balance of £100 left.

3 An **instrument** for weighing or comparing **masses** or **weights**.

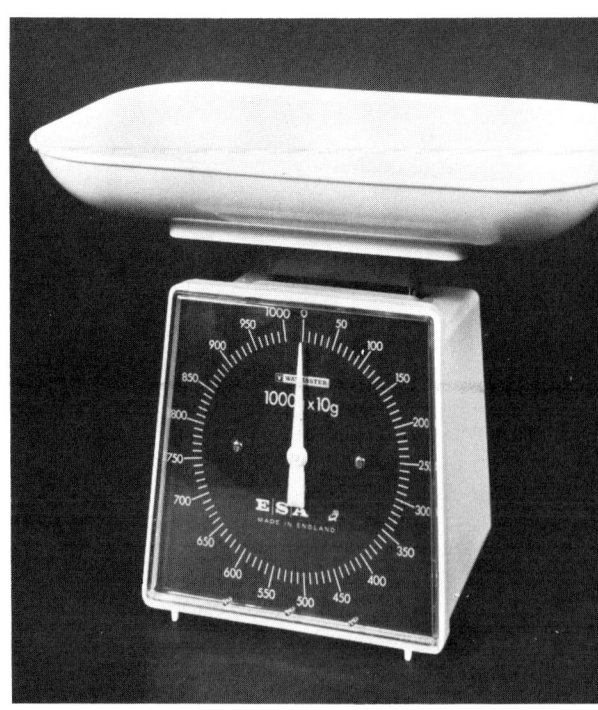

BAR

1 The **line segment** used in a **fraction** to separate the **numerator** from the **denominator**. Generally called a **solidus**.

$$\frac{2}{3} \leftarrow \text{bar or solidus.}$$

2 Bar is used in work on **logarithms**.
$\bar{2}.1$. The bar shows the 2 is **negative**, but the .1 is **positive**, $\bar{2}$ is called a negative characteristic.
$-1.9 = -2 + .1$. This is written as $\bar{2}.1$.

BAR GRAPH

A method of representing information in which **frequencies** are shown by bars of **equal widths**. The **lengths** of the bars are proportional to the frequencies. There are several different interpretations of bar graphs:

1 As in **column graph**. Also called bar chart.

2 When the bars are **horizontal**.

3 When the **columns** are very narrow, even reduced to **lines**.

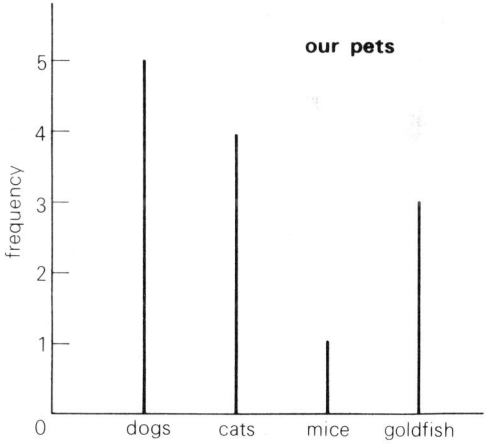

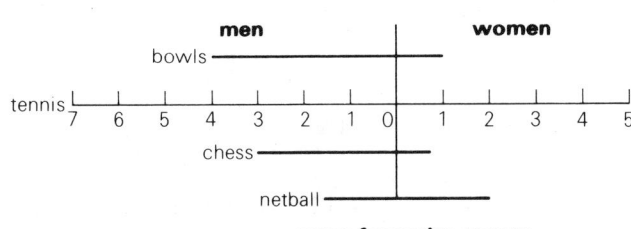

some favourite sports

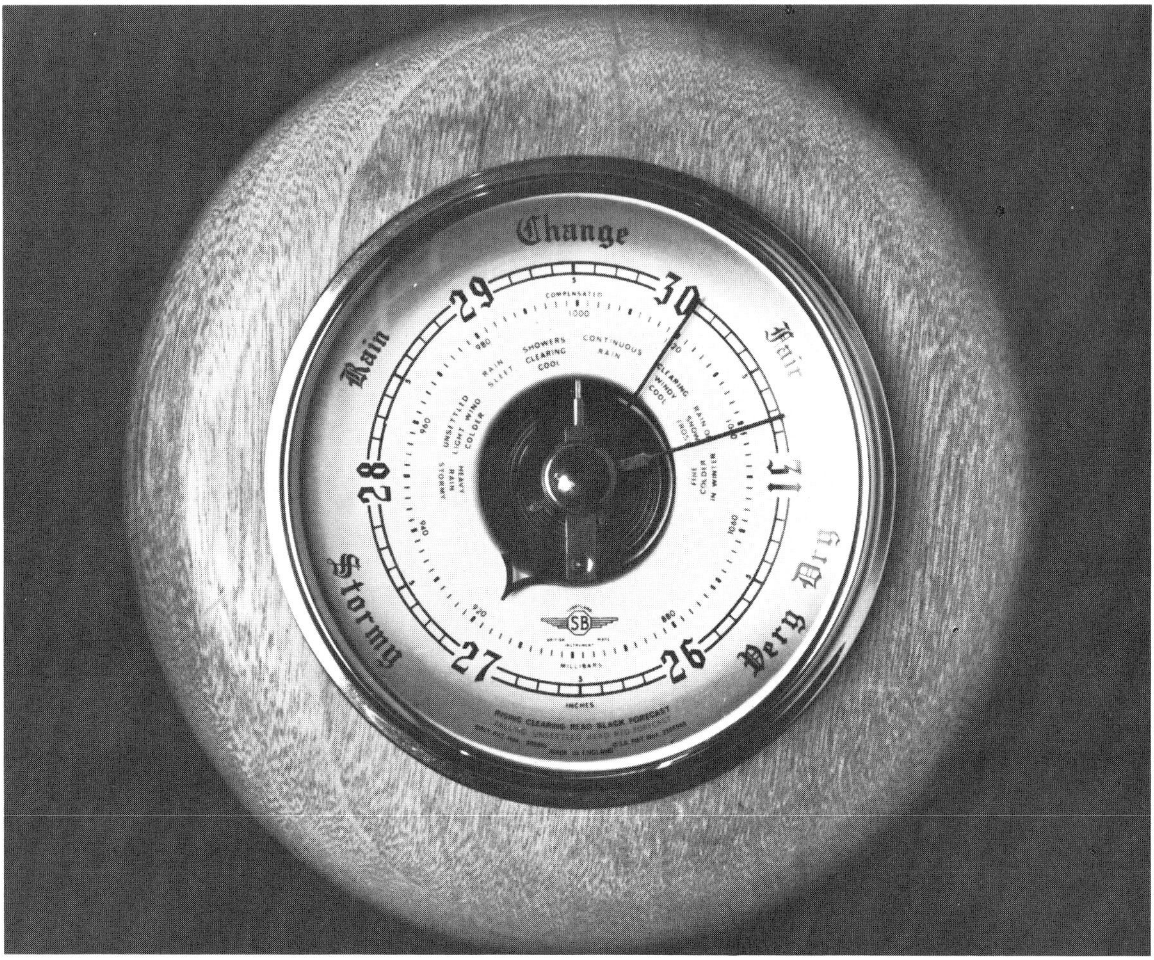

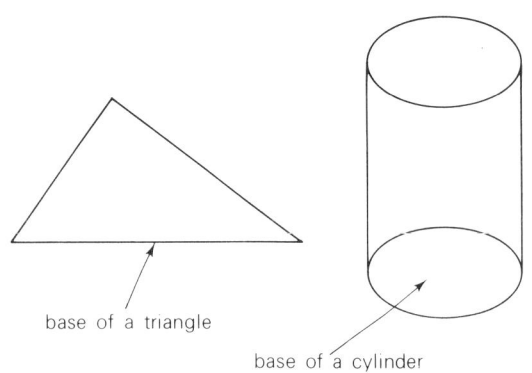

BAROMETER
An **instrument** for measuring atmospheric **pressure**.

BASE
1 That part of a **figure** on which it appears to stand. Bases are generally represented as **horizontal** but need not be so.

base of a triangle

base of a cylinder

Any **side** of a **triangle** can be thought of as a base. The **term** base is especially used when referring to the side to which a **perpendicular** has been drawn.

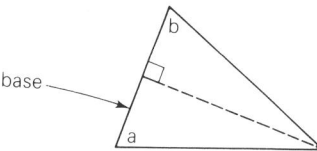

The **angles** marked a and b are called the base angles.

2 The **number** on which a numeration system is founded. We use base ten so that 342 is $3 \times 100 + 4 \times 10 + 2 \times 1$ that is $3 \times 10^2 + 4 \times 10^1 + 2 \times 1$. In base five 342 would be $3 \times 5^2 + 4 \times 5 + 2 \times 1$.

3 In **logarithms**:
$1000 = 10^3$. We say the logarithm of 1000 is 3 with base ten.

$2^4 = 16$ The logarithm of 16 is 4 when the base is 2. This is written as $\log_2 16 = 4$.

BASE LINE
An accurately measured **line** used as a starting line when surveying by **triangulation**.

BC
Abbreviation for 'before Christ'. BC shows that the **number** of years given is before the birth of Christ. Solomon started to build the temple in Jerusalem 961 BC.

BEARING
The **direction** of a **line** with respect to a **North**-South line. A true bearing is the **angle** between the line and a line to the North Pole or to the South Pole. A magnetic bearing is the angle between the line and a line to the magnetic pole. (The magnetic pole is not the same as the North Pole). Whole **circle** bearings are measured **clockwise** from North.

330°

Nautical bearings are measured from North or South according to which results in the smaller **number** of **degrees**.

Example:

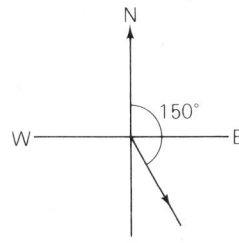

Whole circle bearing 150°
Nautical bearing S 30°E

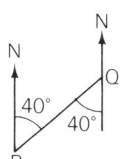

The bearing of Q from P is N 40°E
The bearing of P from Q is S 40°W

The *forward bearing* of Q from P is N 40°E and the *backward bearing* (or back bearing) of Q from P is S 40°W.
OLD ENGLISH *beran*, bring forth.

BI
Bi means two or twice.
Examples: bilateral, two sides; bicycle, two wheels; biped, two feet; **bisect** cut into two **equal** parts or sections. Other words containing bi are biannual, **binary**, **binomial**, **bilateral**.
LATIN *bi* or *bis*, twice.

BILATERAL SYMMETRY
Also called mirror **symmetry**.

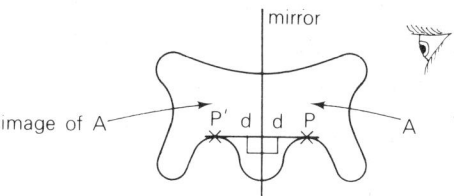

Looking into a mirror from the **right** the **image** of **shape** A is seen in the mirror.
To find the image (P′) of a **point** (P) draw a **line** from P at **right angles** to the mirror. Call the distance from P to the mirror d. The point P′ is a distance d behind the mirror.
LATIN *bi*, twice; *lateris*, a **side**.
GREEK *syn*, together; *metron*, a **measure**.

BILLION
In the United Kingdom and Germany 1 000 000 000 000 (this is 10^{12}).
In the United States and France 1 000 000 000 (this is 10^9).
LATIN *bi* (*bis*), twice; *mille*, a **thousand**.

BINARY
The binary **number** system is based on 2. This system uses only 0 and 1 together with **place value**. **Calculations** in **base** 2 were developed by the German mathematician Leibnitz (1646–1716). Many **computers** use this system, the 0 and 1 corresponding to the electric current being off and on.
Example: 10011_{two} is 19 in the base ten system.

$2^4 = 16$	$2^3 = 8$	$2^2 = 4$	2	1
1	0	0	1	1

$$16 + 0 + 0 + 2 + 1 = 19.$$

LATIN *bi* or *bis*, twice.

BINARY OPERATION
An **operation** on two **members** or **elements** from a **set** which results in a third element which is also in the set.
Example: Consider the operation of **addition** on the set of **natural numbers** (1, 2, 3, 4, . . .). If any two are added the result is a number which is in the set. We say addition is a binary operation on the set of natural numbers. Division is *not* a binary operation on the set of natural numbers because $9 \div 2 = 4\frac{1}{2}$ and $4\frac{1}{2}$ is not in the set.
Some people define a binary operation as *any* operation on two members of a set, even if the result is not in the set. With this interpretation division *is* a binary operation on the set of natural numbers.

BINOMIAL

An expression containing two **terms** that are added or subtracted.

Examples: $5+7$, $x-2$, $5x+2y$, $9x^2-3x$.

LATIN *bi* or *bis*, twice; *nomen*, a name or term.

BISECT

To cut into two **equal** parts.

A **line** cutting another line or **angle** into two equal parts is called a bisector.

Examples:

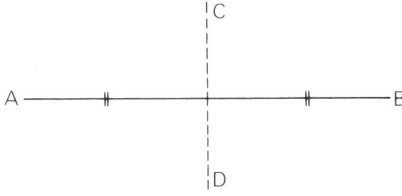

CD is a bisector of AB as it separates it into two equal parts. If it is at **right angles** to the line it is called the **perpendicular bisector**.

BP is the bisector of angle ABC.

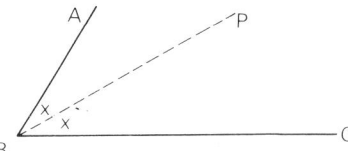

LATIN *bi*, twice; *sectum*, to cut.

BIT

An **abbreviation** for **binary digit**, therefore 0 or 1. (See BINARY.)

BLOCK GRAPH

A **graph** in which the **total numbers** for any one **quantity** are represented by the **area** of a **rectangle**.

If the **measurements** on the **horizontal axis** are **equal**, as in the illustration, then it is called a **column graph**.

The **terms** column graph and block graph are often used to mean the same thing.

BORROW

A **term** used in some methods of **subtraction**.

'Borrow' and 'pay back' should both be avoided because it is not clear who you borrow from or pay back to.

Example:

$$\begin{array}{r} 4^{1}2 \\ -1^{2}9 \\ \hline 2\,3 \end{array}$$

9 from 2, cannot. Borrow 10.
9 from 12, 3.
Pay back. 2 from 4 is 2.

The method would be acceptable if we said 'add ten to each **number** as this does not alter their **difference**'.

BOUNDARY

A **line** that defines the **limits** of a **region**.

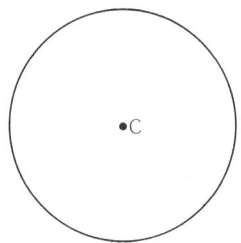

Example: All **points** on the boundary of a **circle** are the same distance from the **centre**, C.

BRACES

braces

These are used to denote a **set**. The **members** or **elements** are contained inside this special form of **brackets**.

Example: The set of vowels can be written $\{a, e, i, o, u\}$.

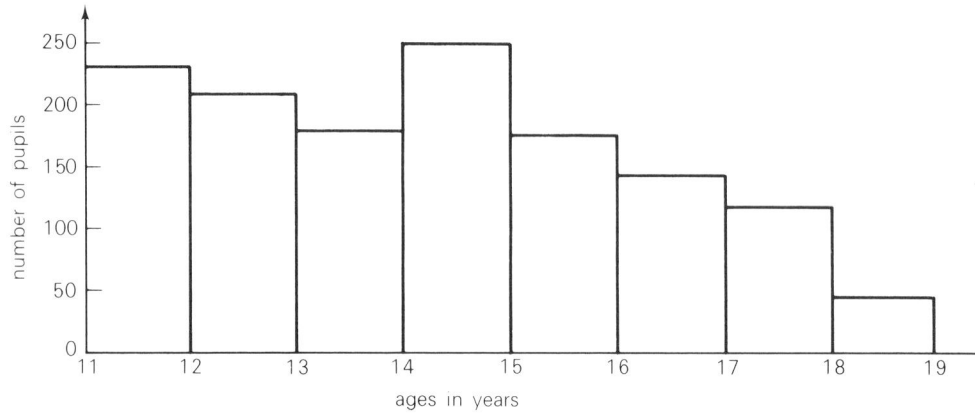

ages of pupils in our school

ages in years

BRACKETS
These may be in a variety of **shapes**.

| square brackets | curly brackets or (braces) | ordinary brackets or (parentheses) |

Brackets show that the **terms** contained by them are to be regarded as together.

Example: $(3 \times 4) + 2 = 12 + 2 = 14$
$3 \times (4 + 2) = 3 \times 6 = 18.$

BREADTH
The distance from one **side** to another. **Width.** When two **measurements** are involved, the shorter is generally, but by no means always, called the width and the longer one the **length**.

Examples:

1 The width of a plank.

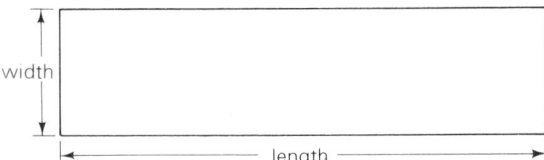

2 The width of a river.

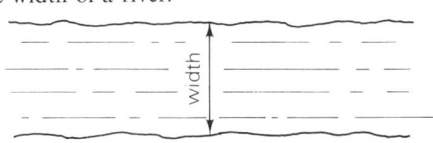

BROAD
Having considerable **width** or **breadth**.

Example: He was broad shouldered. The river was very broad.

BROKEN LINE
1

A **set** of **lines** joined end to end but not in a **straight** line.

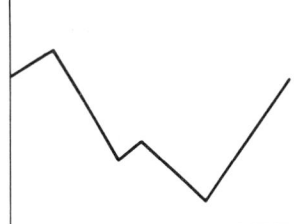

A broken-line **graph**.

2 Sometimes used when 'dashes' are used – – – – – – – but this should be avoided as 1 has important **mathematics** developed from it and confusion may arise.

BUSHEL
A **unit** of **dry measure** used for fruit and grain. It is not now an official unit and is likely to die out altogether.

1 bushel = 8 **gallons** = 4 **pecks**.

C

C

1 The Roman **numeral** for 100. When C is written with a bar above, C̄ it stands for 100 000.

2 C is also used for the following **abbreviations**:

a **circumference**, C = 2πr.

b the **speed** of light.

c **temperature** on the **Celsius** scale, also **Centigrade**.

d a tone in music.

CALCULATION

Any mathematical procedure requiring some thought. This would not, for instance, include straightforward measuring.

LATIN *calculare*, to calculate with the aid of small stones, (*calx*, a stone). The Romans used pebbles or small stones in the same way as beads are used on an **abacus**.

CALCULUS

Any system of **operations** involving the use of **symbols**. There are many forms such as the calculus of **probability** or the calculus of **variations** but it is usually taken to mean the **infinitesimal** calculus.

This was first developed by **Archimedes** but was taken much further by Isaac **Newton** and the German mathematician Leibniz (17th **Century**). Calculus deals with **limits**. when **quantities** get smaller and smaller.

Calculus is of great importance in engineering, physics and other branches of science.

CALENDAR

A **division** of **time** into **years**, **months**, **weeks** and days. It is based on the time of **rotation** of the earth about the sun (year), the moon about the earth (month), and the earth about its **axis** (days).

These do not have a simple connection and modifications are therefore made. Due to **variations** the **average** times are used and a correction made by adding a day in **leap years**.

The **Julian calendar** (46 BC) was introduced by Julius Caesar and the **Gregorian calendar** by Pope Gregory XIII (1582).

CALIPERS

Also spelt callipers.

An **accurate instrument** for measuring or comparing the distance across objects. Calipers can also be used to **measure** thickness. There are many different types. On some a fixed **scale** allows direct readings to be made. On others the distance is transferred to a **ruler** and hence measured.

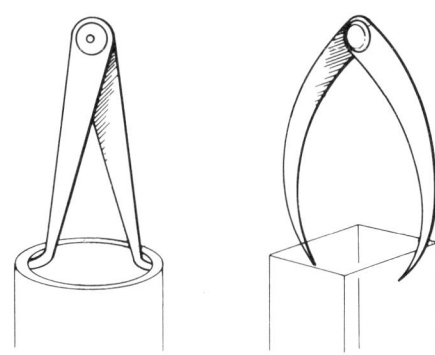

FRENCH *calibre*, the bore of a gun.

CANCEL

1 To **divide** both **numerator** and **denominator** of a **fraction** by a **common factor**.

Example:

$$\frac{24}{30}.$$

Dividing numerator and denominator by 6 gives

$$\frac{2\overset{4}{\cancel{4}}}{\cancel{3}\underset{5}{0}} = \frac{4}{5}.$$

Similarly with **equations** $3x = 12$

$$\frac{3x}{3} = \frac{12}{3} \quad x = 4.$$

2 To **add** or **subtract amounts** to both **sides** of an equation.
Example: $x - 2 = 5$. Add 2 to each side.
$$x - 2 + 2 = 5 + 2. \ x = 7.$$
$$x + 3 = 6,$$
$$x + 3 - 3 = 6 - 3. \ x = 3.$$

CANDLE CLOCK
A device for telling the **time**.

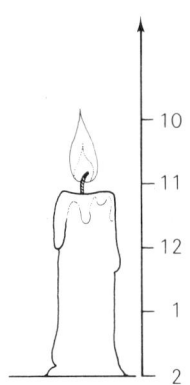

The **scale** enables the time to be read.

CAP
∩ The **symbol** for the **intersection** of **sets**: A ∩ B is read as A cap B or A intersection B.

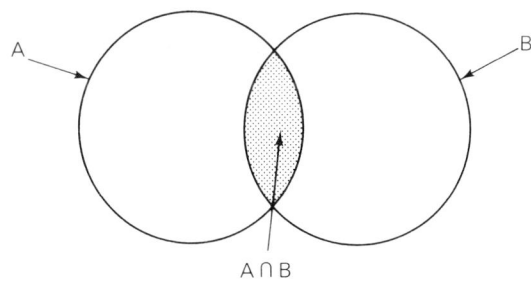

The shaded **area represents** A ∩ B.

CAPACITY
1 The **concept** of **volume** applied to liquids or materials that can be poured. The main **metric unit** of capacity is the **litre** (equivalent to 1000 **cubic** centimetres).
2 'Cubic capacity' is often used with reference to the internal volume and then units for volume are stated and not those of capacity.
Example: The cubic capacity of a car's cylinders is generally given in cubic centimetres (cm^3).

CARAT
A **unit** of **weight** used for gems. Also used as a **measure** of the fineness of gold. 24 carat gold is pure gold but 9 carat contains only $\frac{9}{24}$ gold.

CARDINAL NUMBER
A **number** used to describe how many **members** or **elements** there are in a **set**.
A set has a stated cardinal number n only if its elements can be put in **one-to-one correspondence** with the **natural numbers** $1, 2, 3, 4, 5, 6 \ldots$ n.
Example:

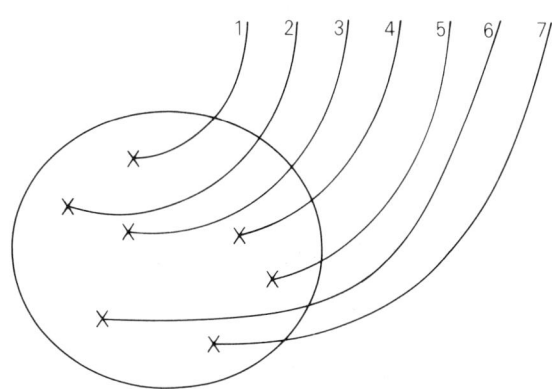

The **empty set** may be said to have the cardinal number 0.
Example: Two sets have the same cardinal number if, and only if, they can be put in one-to-one correspondence.

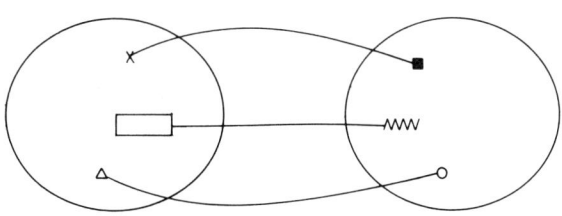

These sets have the same cardinal number 3.
(See ORDINAL NUMBER.)
LATIN *cardinis*, a hinge. (The number system hinges or depends on cardinal numbers.)
GREEK *kardia*, heart. (Cardinal numbers are the heart of our number system, that is, they are very important.)

CARDINAL POINTS

The four main **points** of the **compass**, **north**, south, east and west.

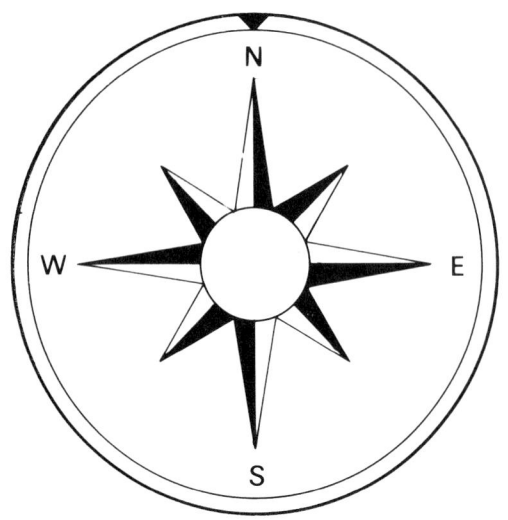

CARDIOID

A **circle** (**centre** A) is fixed and an **equal** circle (centre B) is rolled round it. P is a point on the **circumference** of the second circle.

The path, or **locus** of P is called a cardioid.

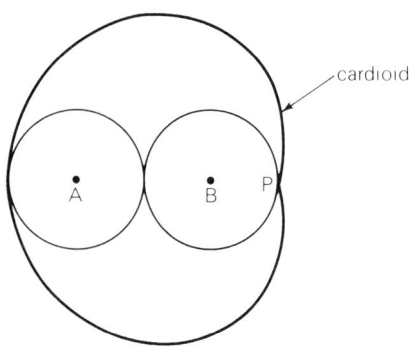

GREEK *kardia*, heart.

CARRY

When adding in our **base** ten system the **total** of the **units**, or ones, may be **greater than** nine. Suppose it is 29. We 'carry' the 2 (tens) as in this example:

```
    18
    37
    25
    19
    ──
    99
```

This is best thought of as regrouping or renaming. The 29 units are renamed as 2 tens and 9 units.

CARTESIAN COORDINATES

A reference system invented by the French mathematician **Descartes** (1596–1650) whereby any **point** in a **plane** is determined by its distance from two fixed **lines** called the axes (see AXIS). These lines are generally at **right angles** and the point where they meet is called the **origin**.

Every point has an **address** which is an **ordered pair**, such as (4, 2) in the example.

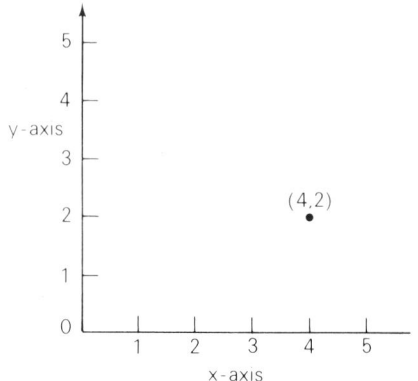

The first **number** (4) gives the distance from the **vertical axis** and the second number (2) gives the distance from the **horizontal** axis. The first number is called the **abscissa** and the second the **ordinate**. Together they are called the coordinates of the point.

Cartesian: Named after **René Descartes** a French philosopher, scientist and mathematician.

CARTESIAN PRODUCT

A **set** of **ordered pairs** formed by matching every **element** in one set with every element in a second set. If the first set is {a, b, c} and the second set {5, 9} then the Cartesian product is {(a, 5), (a, 9), (b, 5), (b, 9), (c, 5), (c,, 9)}. Note that with three elements in one set and two in the other the Cartesian product has six elements. This shows why the term 'product' is used since $3 \times 2 = 6$.

The Cartesian product is also known as the Cartesian set.

CASTING OUT NINES

247 → Adding the **digits**, $2+4+7 = 13.$ $1+3 = 4$
+686   Adding the digits, $6+8+6 = 20.$ $2+0 = 2$ Adding $4+2 = 6.$
──
933
        Adding the digits, $9+3+3 = 15.$ $1+5 = 6 →$ If the **addition** has
                                                    been done correctly
                                                    these two digits will
                                                    be the same.

This method of checking is called CASTING OUT NINES.
There are similar methods for checking **subtraction**,
multiplication and **division**.

                        (Adding digits)
Subtraction   483 → $4+8+3 = 15.$ $1+5 = 6$ Subtract $6-3 = 3$
              −129 → $1+2+9 = 12.$ $1+2 = 3$
              ──
                354 → $3+5+4 = 12.$ $1+2 = 3 →$ These agree.

                        (Adding digits)
Multiplication   29 → $2+9 = 11.$ $1+1 = 2$ ⟍ Multiply
              ×16 → $1+6 = 7.$ $2 \times 7 = 14.$ $1+4 = 5$
              ──
              464 → $4+6+4 = 14.$ $1+4 = 5 →$ These agree.

Division   Change to Multiplication.
          For example $312 \div 12 = 26.$ Instead check whether 12×26 equals 312
Adding digits $1+2 = 3$
                              Multiply, 24. $2+4 = 6$
        $2+6 = 8$

        $3+1+2 = 6 →$ These agree.
(You *cannot* just add the digits and then divide. $3+1+2 = 6.$ $1+2 = 3.$
$6 \div 3 = 2.$ $2+6 = 8$
        These do *not* agree.

CELSIUS

The Celsius **scale** for **temperatures** is named after the
Swedish scientist Celsius (1701–1744). It was formerly
known as the centigrade scale as it has a hundred **units**
between 0° (the freezing **point** of water) and 100° (the
boiling point of water).
The **abbreviation** for **degrees Celsius** is °C.

CENT

A coin in the United States of America and in Canada.
(See PER CENT.)
LATIN *centum*, a hundred.

CENTI

A prefix (written in front of a word) meaning one
hundredth.
Examples: 1 centigram = $\frac{1}{100}$ **gram.**
            1 centilitre = $\frac{1}{100}$ **litre.**
            1 centimetre = $\frac{1}{100}$ **metre.**

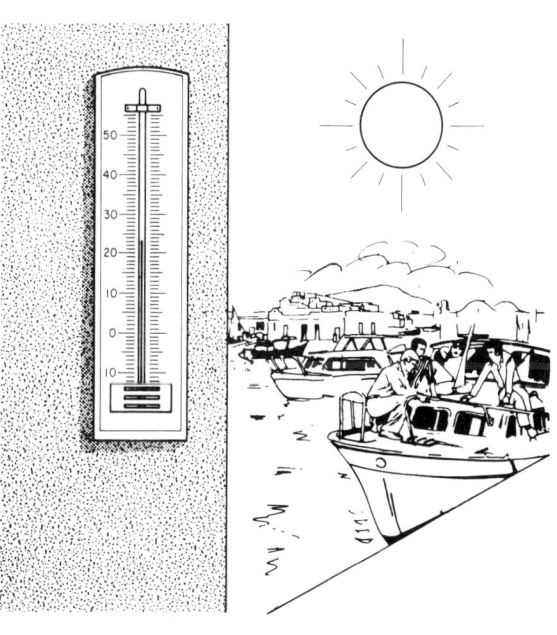

CENTRE

1 The middle **point.** For a **circle** it is the point **equal** distances from the points on the **circumference.**

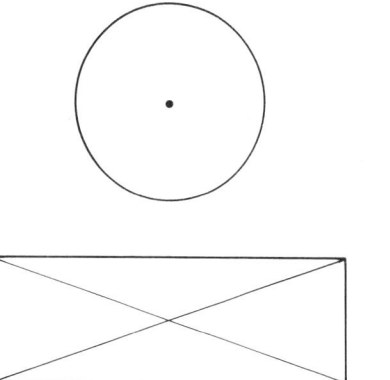

For a **rectangle** it is where the two **diagonals** meet.

2 The centre of **symmetry** is a point such that *any* **line** from it to the **figure** (CP), has a corresponding point on the figure (P′) such that CP = CP′.

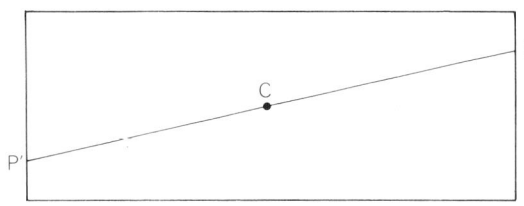

For a rectangle this gives the same point as found in 1.

3 The centre of **gravity.** The point at which the **weight** of a body can be thought to be concentrated since it has the same effect as far as turning and balancing are concerned. (Under most conditions this is the same as centre of **mass.** It would be different if the **density** was not uniform.)

CENTROID

Provided a **figure** is made of the same material throughout the centroid is the same **point** as the centre of gravity. (See CENTRE.)
Example: When the material is not **uniform.**

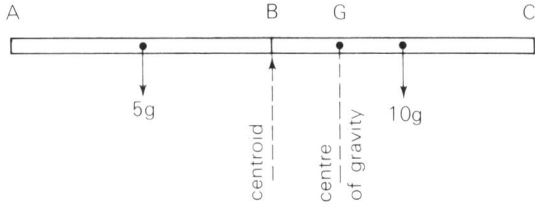

Two strips of **equal length** AB and BC are made of different materials. The centre of **gravity** of AB is at its **mid-point** as shown so its **weight** of 5 g can be taken as acting there. Similarly with the 10 g weight of BC. The centre of gravity of the combined strips is at G. (The weights are in the **ratio** 1 to 2 and the distance from G in the ratio of 2 to 1.) The centroid does not take these **differences** of weight into account and is at B, the mid-point of the combined strip.

CENTURY

A hundred.

A hundred **years,** particularly when reckoned from the accepted year of Christ's birth. We are now in the 20th century. Note that 1066, for example, is in the 11th century, *not* the 10th.

A hundred runs at cricket.

CHAIN

1 A **unit** of **length** based on **Gunter's** chain. 1 chain = 22 **yar**

2 Gunter's or surveyor's chain. A metal chain made from 100 **links.** Edmund Gunter suggested this **unit** of length for **surveying** in 1620. He chose this particular length because 10 **square** chains would then **equal** one **acre.**

1 square chain = 22 × 22 square yards = 484 square yards.
10 square chains = 4840 square yards = 1 acre.
80 chains = 1 **mile.** (22 yards × 80 = 1760 yards.)

CHANCE   See PROBABILITY.

CHANGE

1 The money given to a purchaser being the **difference** between the **amount** given to the shopkeeper and the value of the goods bought.

A book costs 28p. The change from £1 would be 72p.

2 Change of **base.** 25 in base ten is 2 tens and 5 **units** (or ones). Expressed in base seven this becomes 34, that is 3 sevens and 4 units.

There has been a change of base from ten to seven.

This is written as $25_{\text{ten}} = 34_{\text{seven}}$

CHECK

1 The process by which a **calculation** is confirmed as correct – or it is found that an error has been made.
The check could be a repetition of the original calculation but it is better to use a different method.
For example if **columns** are normally added from top to bottom then check by adding from bottom to top.
Multiplication can be checked by **division** and division checked by multiplication.

$$28 \times 3 = 84, \quad 3\overline{)84} \atop 28$$

Similarly with **subtraction**; check by **addition**

$$\begin{array}{r} 84 \\ -69 \\ \hline 15 \end{array} \qquad \begin{array}{r} 69 \\ +15 \\ \hline 84 \end{array}$$

2 The mark $\sqrt{}$ for 'right' is sometimes called a tick, check or check mark.

CHORD

A **line** joining two **points** on a **curve**. Most frequently used when the curve is a **circle**.

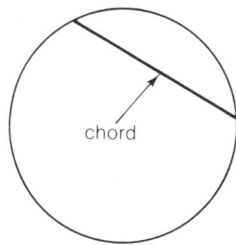

chord

GREEK *chorde*, a string.
A chord resembles a stretched piece of string.

CHRONOMETER

A very **accurate** **clock** designed for use at sea. The exact **time** is required for finding a ship's **position**.

GREEK *chronos*, time,
LATIN *meta*, a **boundary**.

CIPHER

1 A **term** for **zero**.

2 One particular type of **code**. A cipher has a **symbol** for every letter, but a code covers this and many other forms of secret writing.
Example of cipher: A = 1, B = 2, C = 3, D = 4, and so on up to Z = 26.
the becomes 20, 8, 5.
ARABIC *cifr*, zero.

CIRCLE

1 A **set** of **points** in a **plane** that are all the same distance from a fixed point in that plane. This fixed point is called the **centre**.

2 Also used for the **region** inside the set defined in 1. The **boundary** is called the **circumference** but circumference is also used to denote the distance round the boundary. There is thus considerable ambiguity in the use of these **terms**. If 1 is used then the **region** inside the circle is often called a **disc**.

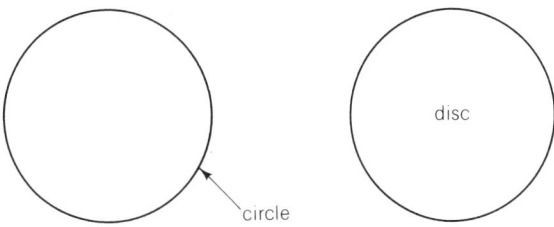

disc

circle

The circumference of a circle is given by $2\pi r$ where r is the **radius**; or πd where d is the **diameter**.
The **area** of circle, or region inside it according to which definition is used, is πr^2 where r is the radius.
(See GREAT CIRCLE.)

CIRCUMFERENCE

The **perimeter** of a **circle**, that is the distance round a circle. If the **radius** is r **units** the circumference is $2\pi r$ or πd units where the **diameter** is d units. Note that circumference is a length and not the **closed curve** itself.

CIRCUMSCRIBED POLYGON

A **polygon** with all its **sides** **tangents** to a **curve**.

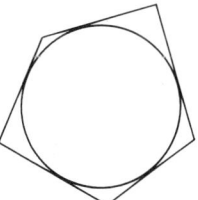 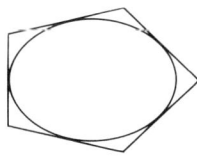

LATIN *circum*, **round**; *scribere*, write.
GREEK *polus*, many; *gonia*, **angle**.

CLASS
See SET.

CLICK WHEEL
See TRUNDLE WHEEL.

Click probably so called due to its sound.
OLD ENGLISH *hweol*, wheel.

CLINOMETER
An **instrument** for measuring the **angle** of **elevation**.

Basically it consists of a means of sighting and a **semicircle** on which a **scale** has been marked in **degrees**. A small **mass** is suspended from string or thread to act as a **plumbline**. The angle shown is 30° in the illustration. It equals the angle of elevation (a in the **diagram**).

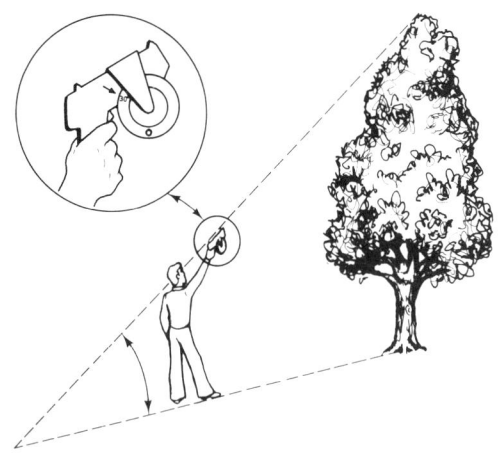

GREEK *klinein*, to lean.
LATIN *meta*, a **boundary**.

CLOCK
A mechanism for measuring **time**.

The picture shows the most famous clock in the world, Big Ben. It is at the Houses of Parliament in London.

CLOCK ARITHMETIC See MODULAR ARITHMETIC.

CLOCKWISE
In the **direction** the hands of a **clock** move.

The opposite direction is called **anti-clockwise** or **counter-clockwise**.

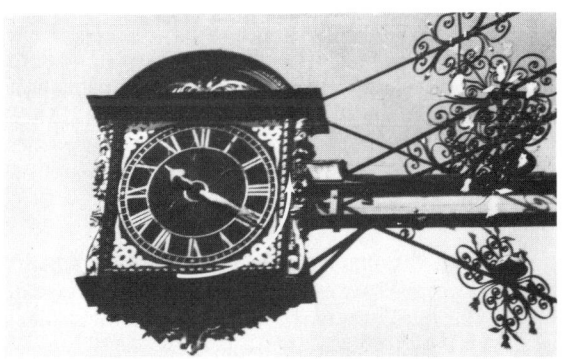

CLOSED CURVE
A **curve** with its ends joined.
If it does not cross itself it is called a simple closed curve.

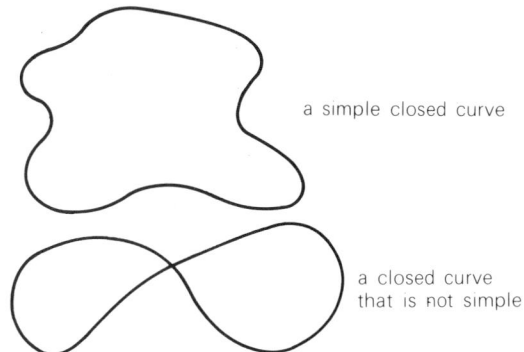

a simple closed curve

a closed curve
that is not simple

The curve and its **interior** are together called a closed **region**.
(See CLOSURE.)

CLOSED SENTENCE
A mathematical **sentence** without a **variable**.
Example: There were six men and nine women. This sentence is closed. 'There were *x* men and *y* women' is not a closed sentence. It is called an **open sentence**.
(See CLOSURE.)

CLOSURE
Two **elements** or **members** of a **set** may be combined by an **operation**. If this results in an element of the set then we have the **property** of closure.
Example: The set of **whole numbers** is closed under **addition**, $3 + 4 = 7$. 3 and 4 are elements of the set and the operation is addition. The result 7 is also a whole number so the set is *closed* under addition. It has the property of closure. This must apply to all elements and not just two as in the example given.
The set of whole numbers is *not* closed under **division**, $3 \div 7 = \frac{3}{7}$ and this is not a whole number.

cm
cm is the **abbreviation** for centimetre or centimetres.

CODE
1 Any method of secret writing. It includes using invisible ink, shorthand, other languages and **ciphers**.
Computers can be used to break codes as they can work through the many possibilities so rapidly.

2 Code is frequently limited to cases where one word, letter or **sign** stands for several others. For example UNT might be agreed to mean 'When is the next meeting?' A cipher has one **symbol** for each letter as in A = 1, B = 2, C = 3, etc.

COEFFICIENT

Generally used to mean the **numerical** coefficient. In $3x$ the x has a coefficient of 3, but we can also regard 3 as having a coefficient of x. More accurately it is the **product** of all the **factors** but with one or more excluded.

For example in $5xy$: 5 is the coefficient of xy, x is the coefficient of $5y$, y is the coefficient of $5x$, $5x$ is the coefficient of y and so on.

COLLECTION

A **term** often used to indicate a **set**. Set is preferred to collection as the meaning of set is more precise.

·COLLINEAR POINTS

Points which lie on the same straight **line**. A, B, C and D are collinear.

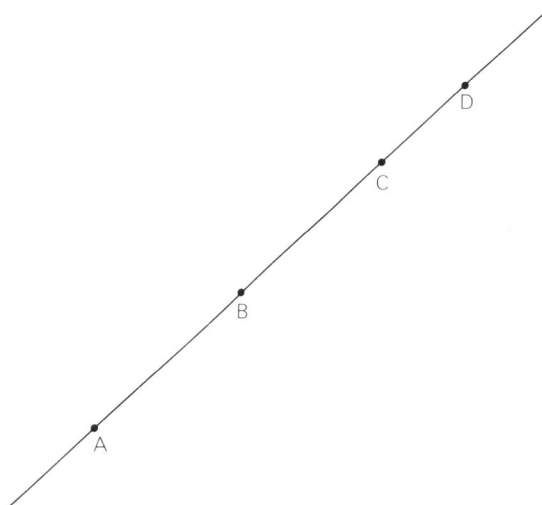

LATIN *co*, with or together; *linum*, flax.

COLOUR FACTOR

A collection of coloured **rods** used to show mathematical **relations** and **operations**, especially **addition**, **subtraction**, **multiplication** and **division**.

These can be done without **number** in the first instance, then later with numbers. The rods are used in a similar way to the **Cuisenaire rods**. Colour Factor has twelve differently coloured rods whereas Cuisenaire has only ten.

COLUMN

A **vertical** arrangement

Example: A column of **numbers**:
12
14
9
8

Compare this with a **row** which is a **horizontal** arrangement such as 12, 14, 9, 8.

The picture shows the most famous column in Britain, Nelson's Column in Trafalgar Square, London.
LATIN *columna*, high.

COLUMN GRAPH

A means of **representation** in which **rectangles** or bars are used to show two **variables**.

In a **bar graph** (or **bar chart**) the bars may be **horizontal** or **vertical** but in a column graph they are vertical.

The intervals on the horizontal **axis** are **equal**. When they are not equal it is called a **block graph**.

The **terms** column graph and block graph are however often used to mean the same thing.
LATIN *columna*, high
GREEK *graphe*, something written.

COMBINATION

A selection of **elements** from a **set** without attention to the **order** of the elements chosen.

Example: When choosing two children from Alan, Betty and Charles there are the following possible pairs. Alan and Betty, Alan and Charles, Betty and Charles. These children in any pairs could just as well have been written in the opposite order, such as Betty and Alan instead of Alan and Betty. Note there are three combinations when selecting any two from three.

If the order was important there would be six possible pairs and we then have a **permutation**. The pairs are (Alan, Betty), (Betty, Alan), (Alan, Charles), (Charles, Alan), (Betty, Charles), (Charles, Betty).

COMMON DENOMINATOR

A **number** which is a **multiple** of the **denominators** of two or more **fractions**.

Example: $\dfrac{2}{3} \dfrac{1}{4}$.

The denominators are 3 and 4.
12 is a common multiple of 3 and 4 and is therefore a common denominator.
By finding a common denominator we can **add** or **subtract** fractions.

$$\frac{2}{3} = \frac{8}{12} \quad \frac{1}{4} = \frac{3}{12}$$

12 is a common denominator.

$$\frac{2}{3} + \frac{1}{4} = \frac{8}{12} + \frac{3}{12} = \frac{11}{12}.$$

24, 36 and 48 are also common denominators of 3 and 4.
12 is in fact the **lowest** (or least) **common denominator** (L.C.D.).

COMMON DIFFERENCE

(Also see **difference**.)
The difference between one **term** and the next in an **arithmetic progression**.
Example: 3, 8, 13, 18, 23, 28, 33, 38, – – – – – – – – .
The common difference is 5.

COMMON DIVISOR

A **number** that **divides** into each of two or more other numbers without leaving a **remainder**.
Example: 3 is a common divisor of 12, 18 and 33. 3 is also said to be a **common factor** of 12, 18 and 33.

COMMON FACTOR

A **factor** of two or more **numbers**. 4 and 6 are common factors of 24 and 36. Also called **common divisors**.

COMMON FRACTION

A **fraction** in which the **numerator** and **denominator** are both **integers**.
$(-3, -2, -1, 0, 1, 2, 3, 4, $ – – – – – – – – are integers.)

Example: $\dfrac{6}{11}$ $\begin{array}{l}\leftarrow \text{numerator} \\ \leftarrow \text{denominator}\end{array}$

Also called **simple fractions**, **vulgar fractions** or **rational numbers**.

COMMON MULTIPLE

A **multiple** of each of several **numbers**.
Example: 18 is a multiple of 2 and also of 3. It is therefore a common multiple of 2 and 3.
(See LOWEST COMMON MULTIPLE.)

COMMON PROPERTY OR ATTRIBUTE

Beads may vary in **shape**, **size**, colour, material and other ways. Any way in which things are the same is said to be a **common property** (or **attribute**).
Examples:

These beads are the same shape and colour but differ in size.
A common property of **rectangles** is that they all have four **right angles**.

COMMUTATIVE PROPERTY OR LAW

This states that the **order** of an **operation** does not affect the result.
Examples: **Addition** of **numbers** is commutative
$3 + 8 = 8 + 3 = 11$.
Multiplication of numbers is also commutative
$4 \times 5 = 5 \times 4$.
$8 - 3$ is 5 but $3 - 8$ is -5 so **subtraction** is not commutative. **Division** is not commutative: $8 \div 2 = 4$ but $2 \div 8 = \frac{1}{4}$.
Washing and then ironing your clothes is not the same as ironing and then washing them (not commutative).
To put on your hat and then your coat has the same result as first putting on your coat and then your hat (commutative).

COMPARE/COMPARISON

1 **Numbers** can be compared in several ways, for example how much greater (by **subtraction**) and how many times greater (**division** or **multiplication**).
Example: Comparing 10 and 2 we can say 10 is 8 **greater than** (or more than) 2 or 10 is 5 times greater than 2.

2 Some **properties** can be compared. Comparison is an essential preliminary to **measurement**.

compare heights   compare areas

 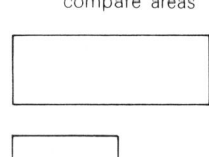

COMPASS

An **instrument** for finding **direction** with respect to magnetic **north.**

A magnetised needle turns on a pivot and **points** towards the magnetic pole. This pole is not the same as the true north pole.

FRENCH *compas*, a **circle**.

COMPASSES

An **instrument** for drawing a **circle** or a circular **arc**. Also used for marking a **point** on a **line** or **curve** at a given distance from a fixed point.

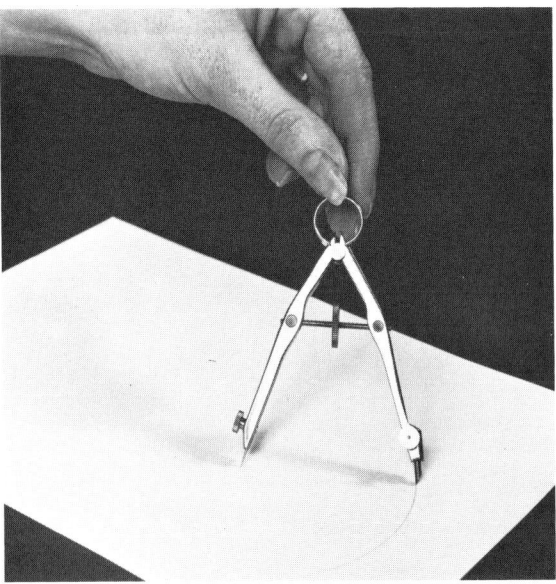

COMPLEMENT/COMPLEMENTARY

1 Complementary **angles**. Two angles whose **sum** is 90°. Each is said to be the complement of the other.

Example :   20° is the complement of 70°.
70° and 20° are complementary.

2 Complement of a **set**.

Let A be the **Universal set** (that is all the **elements** we are concerned with at the moment). B is a **subset** of A. The complement of B is the set of elements that are in A but *not* in B. It is written as B′.

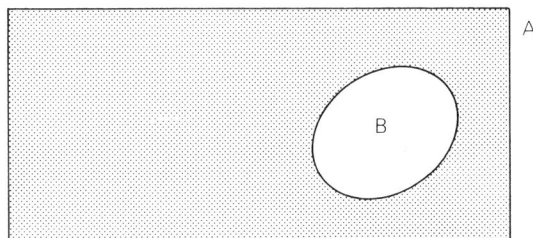

The shaded area represents the complement of B.

3 Complementary **addition**. (**Shopkeeper's method**).

$$\begin{array}{r} 38 \\ -29 \\ \hline 9 \end{array}$$

38–29 is evaluated by finding what must be added to 29 to give 38.

COMPLEX NUMBER

A **number** of the form a + ib where a and b are **real numbers** and i $= \sqrt{-1}$ (or $i^2 = -1$)

Example :   4+i7 or 4+7i.
4 is the real part and i7 (or 7i) the imaginary part.

COMPOSITE NUMBER

A **whole number** which is not **prime**. It therefore has more than two **factors**.

Example : 10 (with factors 1, 2, 5 and 10). Since 7 has only two factors, 1 and 7, it is prime and *not* composite.

COMPOUND INTEREST

Interest paid on the money invested (the **principal**) and also on earlier interest.

Example : £100 principal invested for 2 **years** at 10%. After one year the **amount** (principal plus interest) is £110. After the second year the interest on this is £11 and the amount is £121. The total interest is therefore £21.

When interest is only given on the original amount it is called **simple interest**. The simple interest in the example above would be: (10% of £100) × 2, that is £20.

COMPOUND PRACTICE

See PRACTICE.

COMPOUND PROPORTION

A **quantity** which depends for its **value** on several others. *Example:* The cost of a car depends on the cost of raw materials, wages, as well as other expenses. The income from money depends upon the **amount** you invest, the **length** of **time** of the investment and the **rate per cent** paid in **interest**.

COMPOUND QUANTITY

A **quantity** expressed in more than one **unit**. *Example:* 3 m 24 cm is a compound quantity but 3.24 m is not.

COMPUTATION

The use of mathematical processes to obtain **numerical** results.

These processes may be done mentally or with the aid of an **abacus**, **slide rule**, **computer** or some other device.

COMPUTER

An information processing device. **Slide rules** and **desk calculators** are examples of mechanical computers.

There are two electronic types **analogue** and **digital computers**. An analogue computer is used to copy various conditions and may give a picture or model as opposed to **numbers**.

Digital computers use numbers when calculating generally 0 and 1 (the **binary** system). They print out the answers in **base** ten, the **denary** system.

LATIN *computare*, to reckon.

CONCAVE

Curved inwards. In the example the other **side** is **convex** (curved outwards).

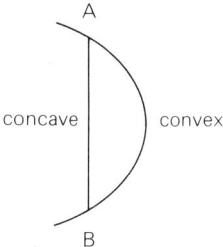

To find if a **curve** is concave join two **points** on the curve (A and B). If the **chord** A B lies entirely on one side of the curve that side is concave.

LATIN *cavus*, hollow.

CONCENTRIC

Circles in the same **plane** that have the same **centre**.

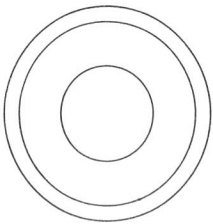

LATIN *con*, with.
GREEK *kentron*, a **point**.

CONCEPT

A mental impression of qualities, **properties**, etc., that is generally represented by a word. An idea and its associations.
Examples: **measurement**, **shape**, truth, . . .

CONCURRENT

Having a **point** in common.

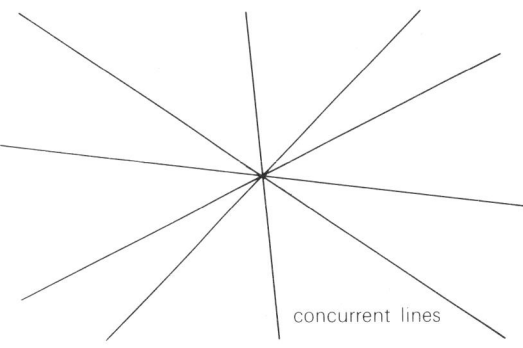

concurrent lines

CONE

A **solid** bounded by a conical **surface** (see below) and a **plane** cutting across it. If this plane cuts the surface in a **circle** a circular cone is formed.

If the **line** from the **vertex** to the **centre** of the **base** is **perpendicular** to the base it is a **right** circular cone.

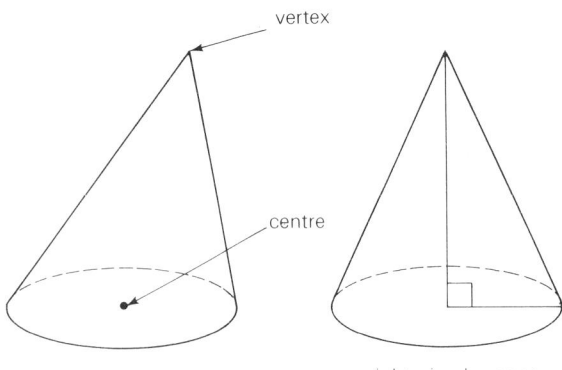

vertex

centre

a right circular cone

Conical surface. The surface formed by lines through a fixed **point** meeting any plane **curve**. This curve is often a circle but need not be.

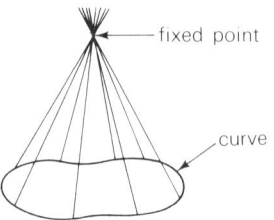

fixed point

curve

GREEK *konos*, cone.

CONGRUENT FIGURES

Those which are **identical** in **shape** and **size**. The **pairs** of **corresponding sides** and **angles** are **equal**.

In modern terms, two figures such that one can be made to coincide with the other by a movement in which the relative **positions** of the parts of the figure are unchanged. Congruent **triangles** are a particularly important instance of congruent figures

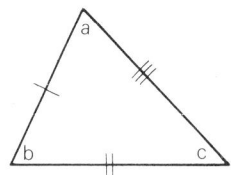
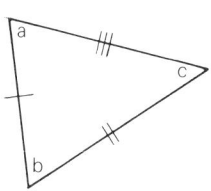

CONGRUENT NUMBERS

Two **numbers** that have the same **remainder** when **divided** by a third number called the modulus. (See MODULAR ARITHMETIC.)
Example: 17 and 41 are congruent, modulus 4, because $17 \div 4$ and $41 \div 4$ both have a remainder of 1.
This is written as $17 \equiv 41$ (mod. 4).

CONIC SECTION

A **curve** formed by the **intersection** of a **plane** with a **right circular cone**.
The curve may be a **parabola**, **hyperbola**, **ellipse** or **circle**.

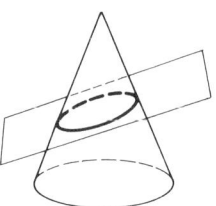

In this case an ellipse is formed.
GREEK *konos*, cone.
LATIN *sectum*, to cut.

CONSECUTIVE
Following in a regular **order**.
Example: 18, 19 and 20 are consecutive **whole numbers**.

CONSERVE/CONSERVATION
To realize which aspects of a situation are not changed.
Example: Conservation of **number**.

Both **sets** have five **members**. The fact that set B is spread out makes some young children think it contains more members than set A. We would then say the child does not conserve number.
Conservation of **length**.
Two pencils are placed together and seen to be the same length. Then one is moved to give the **position** shown:

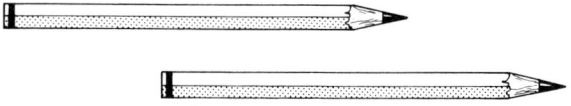

A young child who is not conserving length will say that one pencil is now longer than the other. (See PAIGET.)
LATIN *consevare*, to keep.

CONSTANT
A **quantity** which, under stated conditions, does not vary.
Example: $y = 3x + 7$. The 3 and 7 are constants. The y and x are not. The y and x are **variables**.

CONSTRUCTION
The drawing of a **figure** to fit given specifications. Particularly used for work with a **straight-edge** and **compasses**.
(The straight-edge is normally a **ruler** but the term implies it can be used for drawing **straight lines** but not for measuring **lengths**).
Example: Given the **angle** ABC construct its **bisector**.

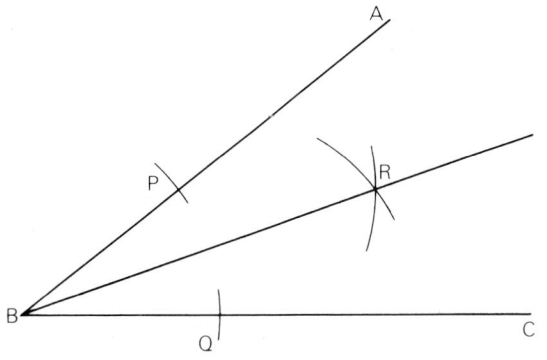

Construction.
Set compasses to any convenient **radius**. With **centre B** cut AB and CB at P and Q. With centres P and Q draw **arcs** to meet at R. Join BR. BR is the bisector of angle ABC.

CONTINUOUS
Without a break.
Example: From A to B the **curve** is continuous.

CONTOUR
A **curve** drawn on a **map** through **points** that are the same **height** above **mean sea-level**.

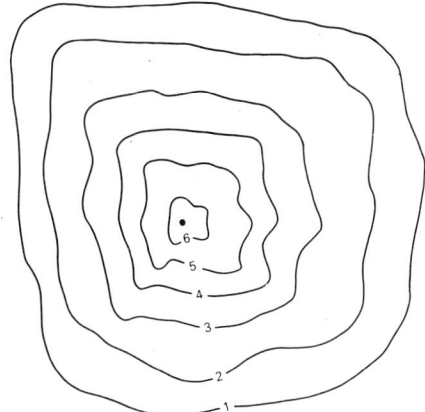

CONVENTIONAL SIGNS
Signs that are generally agreed.
Example: $+$ is the conventional sign for **addition**. Similarly other matters are decided by general agreement. **cm** is the conventional **abbreviation** for centimetre, y^3 is the conventional way of writing $y \times y \times y$.

CONVERGE
To come closer together.
Examples: The three **lines** converge to P.

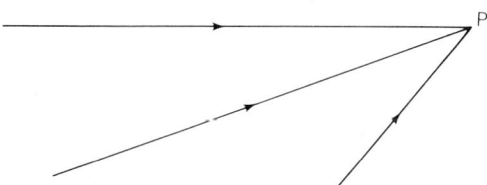

Numbers that tend towards a definite **limit** are called a convergent **sequence**.
$\frac{3}{4}, \frac{4}{5}, \frac{5}{6}, \frac{6}{7}, \frac{7}{8} \ldots$ if these are continued they get nearer and nearer to 1. The number they tend towards is called the limit.

CONVERSE

Stated the other way round. The converse of 'A **square** has four **right angles**' is 'A **figure** with four right angles is a square'. The converse in this case is *not* true for the figure could be a **rectangle** with unequal **sides**.

The converse of 'A figure with three sides has three angles' is 'A figure with three angles has three sides'.

The converse is true.

CONVERT

To change. To convert **metres** into centimetres **multiply** by 100.

16 in **base** ten is 31 when converted into base five. (See CHANGE for change of base.)

A conversion **table** enables us to read off one **measurement** in terms of another.

CONVEX

Curved outwards.

That part of a **curve** or curved **surface** on which the **tangent** lies.

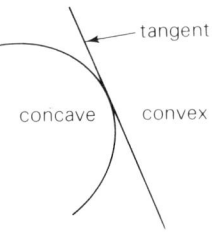

The **reverse** of **concave**.

LATIN *convexus*, from *convehere*, to carry.

Weights and Measures

Metric Measures and Equivalents

Length

1 millimetre (mm)		= 0·0394 in
1 centimetre (cm)	= 10 mm	= 0·3937 in
1 metre (m)	= 100 cm	= 1·0936 yds
1 kilometre (km)	= 1000 m	= 0·6214 mile

Surface or Area

1 sq cm (cm²)	= 100 mm²	= 0·1550 sq in
1 sq metre (m²)	= 10,000 cm²	= 1·1960 sq yds
1 are (a)	= 100 m²	= 119·60 sq yds
1 hectare (ha)	= 100 ares	= 2·4711 acres
1 sq km (km²)	= 100 hectares	= 0·3861 sq mile

Capacity

1 cu cm (cm³)		= 0·0610 cu in
1 cu decimetre (dm³)	= 1000 cm³	= 0·0351 cu ft
1 cu metre (m³)	= 1000 dm³	= 1·3080 cu yds
1 litre (l)	= 1 dm³	= 0·2200 gallon
1 hectolitre (hl)	= 100 litres	= 2·7497 bushels

Weight

1 milligramme (mg)		– 0·0154 grain
1 gramme (g)	= 1000 mg	= 0·0353 oz
1 kilogramme (kg)	= 1000 g	= 2·2046 lb
1 tonne (t)	= 1000 kg	= 0·9842 ton

British Measures and Equivalents

Length

1 inch		= 2·54 cm
1 foot	= 12 inches	= 0·3048 m
1 yard	= 3 feet	= 0·9144 m
1 rod	= 5·5 yards	= 5·0292 m
1 chain	= 22 yards	= 20·117 m
1 furlong	= 220 yards	= 201·17 m
1 mile	= 1760 yards	= 1·6093 km
1 nautical mile	= 6080 feet	= 1·8532 km

Surface or Area

1 sq inch		= 6·4516 cm²
1 sq foot	= 144 sq inches	= 0·0929 m²
1 sq yard	= 9 sq feet	= 0·8361 m²
1 acre	= 4840 sq yards	= 4046·9 m²
1 sq mile	= 640 acres	= 259·0 hectares

Capacity

1 cu inch		= 16·387 cm³
1 cu foot	= 1728 cu inches	= 0·0283 m³
1 cu yard	= 27 cu feet	= 0·7646 m³
1 pint	= 4 gills	= 0·5683 litres
1 quart	= 2 pints	= 1·1365 litres
1 gallon	= 8 pints	= 4·5461 litres
1 bushel	= 8 gallons	= 36·369 litres
Apothecaries		
1 fluid ounce	= 8 fl drachms	= 28·413 cm³
1 pint	= 20 fl ounces	= 568·26 cm³

Weight

Avoirdupois		
1 ounce	– 437·5 grains	= 28·350 g
1 pound	= 16 ounces	= 0·4536 kg
1 stone	= 14 pounds	= 6·3503 kg
1 hundredweight	= 112 pounds	= 50·802 kg
1 ton	= 20 cwt	= 1·0161 tonnes

A conversion **graph** is another means of doing this.

COORDINATES

An **ordered pair** of **numbers** which describe the **position** of a **point** with reference to **lines** (called **axes**). An ordered pair is one in which the order of the numbers is important: (3, 2) is not the same as (2, 3).

Example:

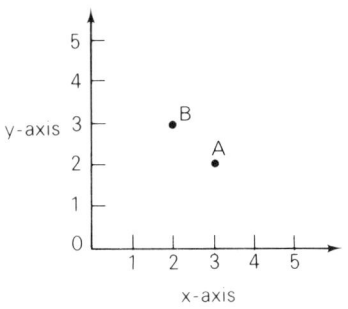

A is the point 3, 2.
B is the point 2, 3.

The first number in a pair shows the distance from the *y*-axis (**vertical** axis) and the second number shows the distance from the *x*-axis (**horizontal** axis).
(In more advanced work on **three dimensional** coordinates three ordered numbers are needed to describe the position of a point).
(See ABSCISSA, AXIS, ORDINATE.)

CORNER

1 A **point** where two **lines** meet. More correctly called **vertex** (plural: vertices).
Example: A **rectangle** has four corners or vertices.

2 A point where several **planes** meet. The correct **term** (as in 1) is vertex.
Example: Three planes meet at the corner of a room.

3 The lines that meet at a point together with the **region** between these lines and close to the point.

The shaded region show the four corners. The term corner is not clearly defined when used in this way. In the example the shaded region could be smaller than shown and yet still be the corner.

CORRESPONDENCE

The **relation** between the **elements** (or **members**) of one **set** and the elements of another set or the relation between some elements of a set and elements in that same set. Correspondence is sometimes used to mean the same as **mapping**.
See RELATION and MAPPING as different people use these terms in different ways.
Example of correspondence:

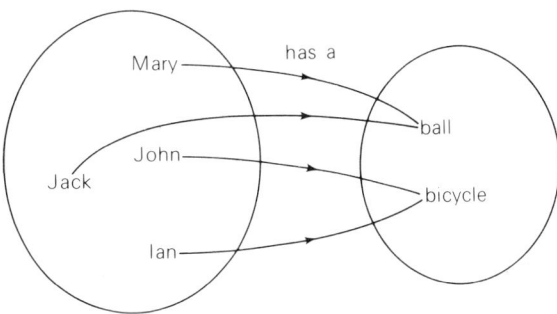

CORRESPONDING

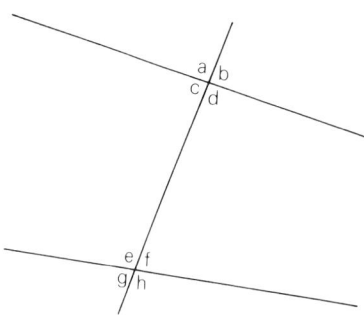

Corresponding **angles**. The angle a and e are in the same relative **position** (top and left) with reference to the **transversal** and the **lines** it cuts. Similarly the **pairs** c and g, b and f, d and h are corresponding angles.

If the lines cut by the transversal are **parallel** then the corresponding angles are **equal**.

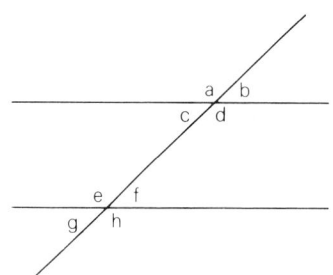

a = e, b = f, c = g and d = h.

Corresponding angles of a **triangle**.

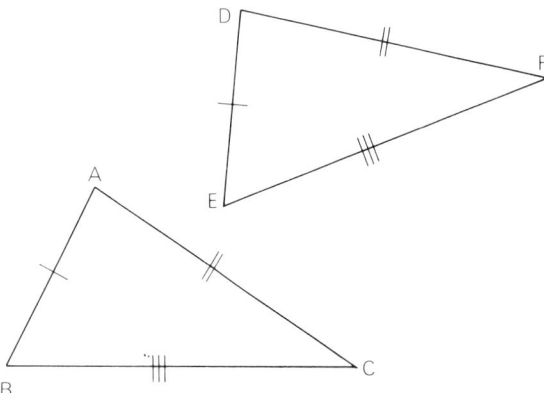

If two triangles are **congruent** the angles in the same relative positions are equal and are the corresponding angles of the triangles; A corresponds to D, B to E and C to F. Similarly we have corresponding **sides**; AB corresponds to DE, BC to EF and AC to DF. A and D, B and E, C and F are corresponding **vertices**.

The same applies if two triangles are **similar**. Other shapes that are similar also have corresponding angles, sides and vertices.

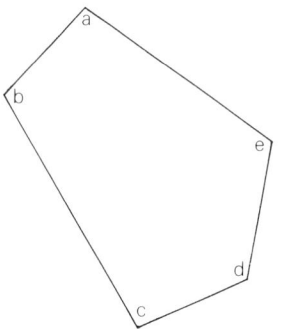 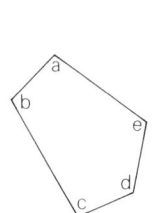

The corresponding angles are equal. The lengths of corresponding sides are in **proportion** (2 to 1).

COST PRICE

The cost of an article to the person buying it. A manufacturer sells goods to a shopkeeper or retailer. The manufacturer's **selling price** is the shopkeeper's cost price. The shopkeeper then sells the goods to a customer. The shopkeeper's selling price is the customer's cost price.

Example: A shopkeeper buys goods for £40 (Cost Price) from a manufacturer and sells them for £50 (Selling Price). His **profit** is £10. (Profit = Selling Price − Cost Price). The customer's Cost Price is £50. Cost Price and Selling Price are abbreviated as C.P. and S.P.

COUNT

To say the **numerals** in order, starting at 'one' and relating each numeral to one, and only one object. Counting involves a **one-to-one correspondence** between the numerals and the objects counted.

$$
\begin{array}{cccccc}
0 & 0 & 0 & 0 & 0 & 0 \\
\uparrow & \uparrow & \uparrow & \uparrow & \uparrow & \uparrow \\
\text{one} & \text{two} & \text{three} & \text{four} & \text{five} & \text{six}
\end{array}
$$

The last mentioned numeral gives the **cardinal number** of the **set**. In the above example this is six. Counting could therefore be described as 'the process of finding the cardinal number of a set'.

COUNTER-CLOCKWISE

See ANTI-CLOCKWISE.

COUNTING NUMBERS

The **set** of **numbers** 1, 2, 3, 4, 5, 6, 7, . . . etc.

COUPLE

1 In **arithmetic**: Another word for two.

2 In **mechanics**: Two **equal parallel forces**, not in the same **straight line**, and acting in opposite **directions**.

CREDIT

1 The extent to which a person can buy or **borrow** goods on trust.

2 The **amount** of money in a person's **account**. Or, the **total** amount of goods and money belonging to a person.

CRITERION

A test or standard that is acceptable.
Example: The criterion for deciding if a knitting needle is an acceptable **length** might be that it does not differ by more than 1 **millimetre** from the length stated on the packet.

CROSS MULTIPLICATION

The process in which two **equal fractions** are replaced by an **equivalent statement** involving **multiplication**.

For instance $\frac{3}{4} = \frac{6}{8}$ becomes $3 \times 8 = 4 \times 6$.

More generally $\frac{a}{b} = \frac{c}{d}$ becomes $a \times d = b \times c$.

The diagram  shows why 'cross' is used.

CROSS PRODUCT

An **operation** on **sets**, generally called the **Cartesian product**.
Also a **term** for one sort of **multiplication** of **vectors**, also known as the vector product.

CROSS-SECTION

The **figure** made when a **solid** is cut by a **plane**.
Example:
1 A **cylinder** cut as shown has a circular cross-section.

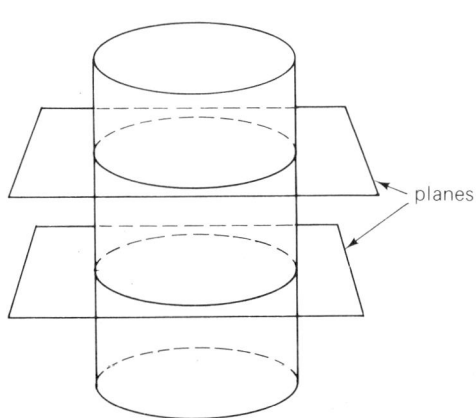

As all the plane cuts **parallel** to the one shown give the **same** figure there is a **uniform** cross-section.

2 Plane cuts parallel to the circular **base** of the **cone** are all **circles** but they all differ in **size**.

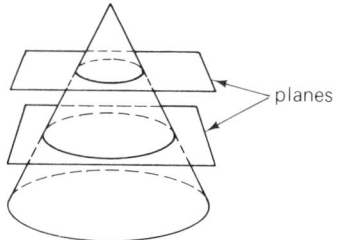

The cone does not, therefore, have a uniform cross-section.
LATIN *crux*, cross; *sectum*, to cut.

CUBE

1 A **regular solid** with six **faces** that are **equal squares**. Each **edge** is **perpendicular** to its adjoining edges.

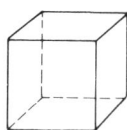

2 In **algebra** the third **power**. $4 \times 4 \times 4$ is written as 4^3 and read as 'four cubed' or 'four to the power of three'.

CUBIC MEASURE

A **measure** of **volume**.
Example: Cubic centimetre. The volume **equivalent** to that of a cube with sides of 1 centimetre. One cubic centimetre is written $1\,cm^3$.

CUBIT

An ancient **measure** of **length**.
The distance from the elbow to the tip of the middle finger.

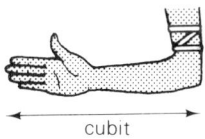

cubit

As it varied from person to person it was not a very suitable **unit** of **measurement**.
LATIN *cubitum*, the elbow.

CUBOID

A **solid** with six **faces** that are **rectangles**, the opposite faces being **congruent** (that is **equal** in all respects). Each **edge** is **perpendicular** to its adjoining edges.

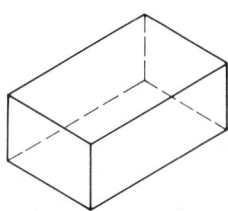

CUISENAIRE RODS

A **set** of coloured **rods** with **lengths** from 1 **cm** to 10 cm and square ends with edges 1 cm. They are named after a Belgian schoolmaster Georges Cuisenaire, who invented them. Rods of **equal** lengths have the same colour.

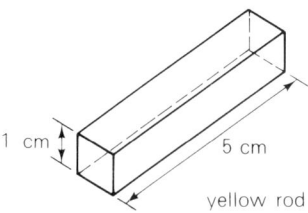

By using the rods children build **relations** and do **operations** that give them a sound base on which later **number** work can be built.
Example:

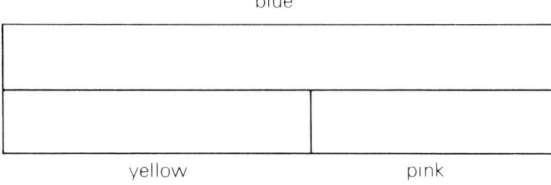

yellow + pink = blue.
blue − yellow = pink.
blue − pink = yellow.
The lengths are yellow 5 cm, pink 4 cm, blue 9 cm so the given example leads to $5 + 4 = 9, 9 - 5 = 4$ and $9 - 4 = 5$.

CUNEIFORM

Wedge-shaped.
The **term** applies especially to writing with a stick (called a stylus) that had a wedge-shaped end. This was used to make marks in clay when writing.
These are examples of **numerals** used in the **Babylonian System**.

1

CUP

Used to denote the **union** of two **sets**. Denoted by the symbol ∪. A ∪ B is read as A cup B or A union B.

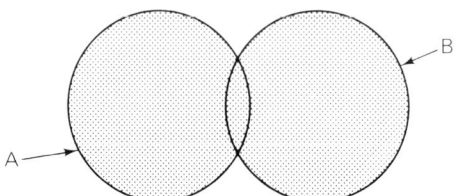

The shaded **region** represents A ∪ B, called C ∪ P to remind us which way up it is. It resembles a cup without a handle. ∩ is called C A P and shows the **intersection** of two sets.

CURRENCY

Money or other materials used when buying or exchanging.
LATIN *currere*, to run. Run is used here in the sense of 'flow'. Currency flows round or circulates.

CURVE

A **line** which has no **straight** part. An **open curve** has end-points. A **closed curve** has no end-points.
Examples:

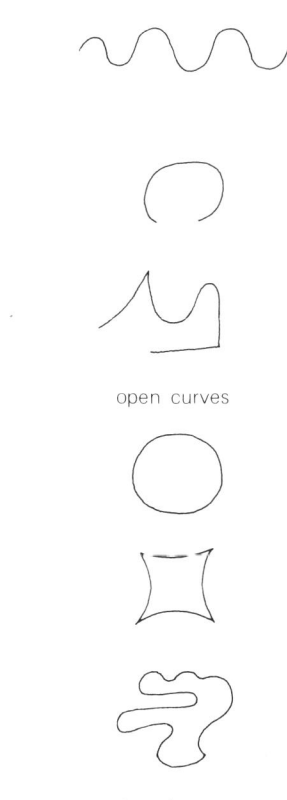

open curves

closed curves

Straight lines could be regarded as special cases of curves. If a circle has a very large **radius** a small **arc** will be very nearly straight.

CYCLE

1 A **sequence** of events that repeat regularly, or

2 The **time** between such repetitions.

CYCLIC

Moving or arranged in a **cycle**.

CYCLOID

The path traced out by **point** (P) on the rim of a **circle** as it **rolls** along a **straight line**.

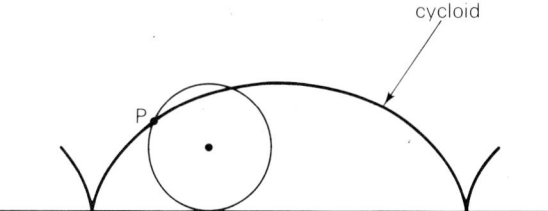

CYLINDER

A **solid** with a **uniform** circular **cross-section**. If the ends are **perpendicular** to the curved **surface** it is a **right** circular cylinder.

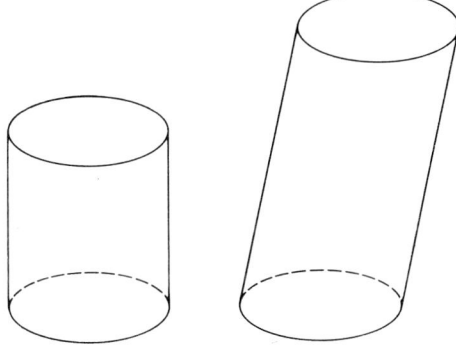

A right circular   A cylinder that is not
cylinder.   right circular.

Examples: A tin of beans, drain pipe.
GREEK *kylindros*, roller.

D

D

The Roman **numeral** for 500.

The Roman **symbol** for 1000 was M, written as ⊂IƆ. D originated from the right half of this.

DATA

Facts or information that is given or collected.

DATE LINE

A **line** on the Earth running mainly along the 180th **meridian**.

By international agreement if this line is crossed in a westerly **direction** one day is added to the date and if crossed in an easterly direction one day is subtracted from the date.

Suppose the sun rose at 6 A.M. one Monday morning and you travelled in an aircraft that followed the Sun in its apparent path round the earth. At all points on your journey you would see the Sun rising and you would be back at your starting point one day later, that is at 6 A.M. on Tuesday morning. Yet it would seem to you that it has always been the dawn on Monday and a day appears to have been lost. When did it stop being Monday and start to be Tuesday?

The **dateline** has been internationally agreed so as to overcome this problem. It passes through the Pacific so that as small a number of people as possible are affected. As your plane crosses the line from East to West the day is changed from Monday to Tuesday. If crossing the line from West to East the day changes from Tuesday to Monday.

In the same way other days change to the one before or the one after according to the **direction** of travel.

DATUM

1 The singular of **data**.

2 The **heights** of **points** on the earth are measured from a stated **level** which is called the datum. For example the Ordnance Survey measures heights from the **mean sea level** at Liverpool.

DEBIT

A **term** in book-keeping showing the **cost** of an item that still has to be paid for.

DECA

Prefix meaning ten.

GREEK *deka*, ten.

DECADE

Ten **years**.

DECAGON

A **polygon** with ten **sides**.

DECAGRAM

Ten **grams**.

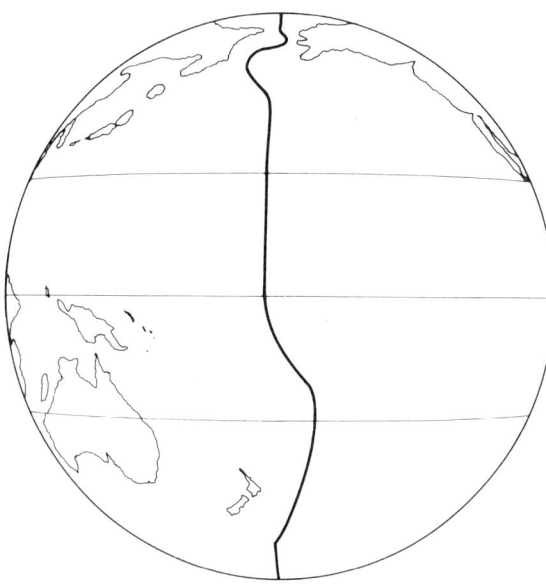

DECAHEDRON
A **solid** with ten **faces**.

DECALITRE
Ten **litres**.

DECAMETRE
Ten **metres**.
(See DEKA.)

DECI
Prefix meaning one-tenth.
LATIN *decimus*, one tenth part.

DECIGRAM
One-tenth of a **gram**.

DECILITRE
One-tenth of a **litre**.

DECIMAL
Connected with, or based upon ten.
Decimal coinage: Coins based on ten, **multiples** of ten and subdivisions of ten.

Decimal **fraction**: A form of fraction in which the **number** of tenths ($\frac{1}{10}$), hundredths ($\frac{1}{100}$), thousandths ($\frac{1}{1000}$), etc. are shown by the position of a **digit** to the right of a **point**, called the decimal point.
Example:
Decimal point: A point . placed after the **units** digit and before the decimal fraction. In the United Kingdom it is written above the **line** (3·48) and on the line (3.48). Printers find it more convenient to place it on the line 3.48. Many countries place the point on the line and some use a comma instead of the point (3,812).

Decimal system: A number system based on ten. Numbers are represented by the digits 0, 1, 2, 3, 4, 5, 6, 7, 8, 9, their **value** depending upon their **position**. (See PLACE VALUE.)
This system includes **whole numbers** and decimal fractions (12.173). The system has been in general use for nearly 400 years.
The Dutch mathematician, Simon Stevin or Stevinus (1584–1620) suggested all **measurements** should be made using the decimal system. He was in charge of the provisions for the army of William of Orange and wrote one of the first books on the **theory** of decimals. He wrote 2.378 as 2⓪3①7②8. Other ways of writing this were 2378 and 2|378. The decimal point, as we know it, first appeared about 1710. The decimal system became very important in Europe when the **metric system** was adopted by the French Assembly during the French Revolution.
LATIN *decimi*, one tenth part.

DECIMETRE
One-tenth of a **metre**.
(See DECIMAL.)

DECOMPOSITION METHOD
A method of **subtraction** in which one ten is 'decomposed' into ten **units**, or one hundred into ten tens and similarly for higher **powers** of ten.
Example:

$$\overset{\scriptstyle 2\ 11\ 1}{3\,2\,4} \\ -\quad 8\,6 \\ \hline \quad 2\,3\,8$$

Main steps.
a One ten is changed to 10 units.
b $14-6 = 8$.
c One hundred is changed to 10 tens.
d $11-8 = 3$. 3 tens.
e There are 2 hundreds.

DEDUCE
To reason or think.
Example: Given that $2x+5 = 13$ we can **deduce** that $x = 4$.

DEDUCT
Take away or **subtract**. Deduct 2 from 10. The **result** is 8. If you misbehave, your parents might deduct 10p from your pocket money.

DEFICIT
The **amount** by which a **quantity** is less than it should be. The **term** particularly applies to money.
Example: On checking the **accounts** a firm found there was a deficit of £800 due to theft.

DEFINITION

A precise description using only **terms** that have already been exactly described (or defined). A definition should have a **minimum** of information.

Example: A **rectangle** is a **parallelogram** with a **right angle**. Note: This definition requires that parallelogram and right angle have already been defined. It is unnecessary to say it has four right angles as this follows from the **properties** of a parallelogram together with the fact that there is one right angle.

[The opposite **angles** of a parallelogram are **equal**, hence there are at least two right angles. Since the **sum** of the angles is 360° and two are right angles the sum of the other two is 180°. These are equal (opposite angles) and therefore each is 90°].

DEGREE

1 A **unit** for measuring **angles**. It is $\frac{1}{360}$ of a complete **rotation**. The choice of 360 probably originated from the **Babylonians** who had 360 days in their **year** and used 60 as their number **base**. One degree is written as 1°.

2 A unit for measuring **temperature**.
(See CELSIUS and FAHRENHEIT.)

3 In **algebra**. x^3 is a **term** of the third degree and so is $4y^3$. x^2y^4 is of the sixth degree and so is $5x^2y^4$. The **power** of the **variables** (x and y in the above cases) are added but the power of any **whole number** is not.
Example: $5^3x^2y^5$ is of the seventh degree. In an **expression** such as $3x^4y + 7y^3 - 2y$ the power of the expression is that of the highest term. In this case the degree is five, that of $3x^4y$.

DEKA

(Usually DECA.) Prefix meaning ten.
Example: dekametre, ten **metres**.
GREEK *deka*, ten.

DENARIUS

An old Roman silver coin. In the New Testament this coin is called a **penny**. A penny in the United Kingdom was therefore abbreviated to d. In 1971 the United Kingdom changed to decimal currency and the **new penny** was denoted by **p**. Coins called dinars in Yugoslavia and Iran have derived their name from denarius.

DENARY SYSTEM

A **number** system based on ten. The number system used in this country is denary. Numbers are expressed as **multiples** of **powers** of 10.
Example: $4327 = 4 \times 10^3 \times + 3 \times 10^2 + 2 \times 10 + 7$.
(As $10^0 = 1$ the 7 could be written as 7×10^0).

The use of ten as our number base probably arose from the fact that we have ten fingers.

DENOMINATOR

That part of a **fraction** which is written below the **solidus**, or **line**.

Example: $\frac{4}{5}$ ← solidus, 5 is the denominator.

The denominator shows how many **equal** parts the **quantity** has been separated into. In the example above there are 5 parts.
LATIN *denominare*, to name.

DENSITY

The **mass** of a **unit volume**. For instance the density of water is approximately 1 **gram** per cubic centimetre.
Example: A mass of 24 g has a volume of 4 cm³. The density is therefore 6 grams per cubic centimetre.

Archimedes discovered some facts about density when he got into his bath and saw that the level of the water rose. He was so excited that he ran out of the bathroom naked shouting 'Eureka' which means 'I have found it'. He had found the solution to a suspected fraud by a goldsmith. King Hiero of Syracuse had asked for a golden crown to be made but the goldsmith had used some cheaper metal hidden beneath the gold. He had been careful to make the crown **weigh** exactly the same as a golden crown would have done. When placed in a bucket full of water the crown displaced more water than an **equal amount** of gold. This showed the volume of the crown was more than it should have been. Since the cheaper metal had a lower density than gold it took up more space.

DEPENDENT

When two **variables** are related one may be selected first and the **value** of the other then depends on this choice. The one selected first is called the **independent** variable and the other is the dependent variable.

Examples: $y = 5x$. The values of the independent variable x are given as 0, 1, 2, 3, 4 and 5. The dependent variable y then has corresponding values of 0, 5, 10, 15, 20 and 25. These values can be recorded as **ordered pairs** (0, 0), (1, 5), (2, 10), (3, 15), (4, 20) and (5, 25). The independent variable is written first and the dependent variable second.

When relating the **height** of a child to its age the height is recorded at **intervals** of six months. The age has been selected (independent variable) and the corresponding height (dependent variable) is then found.

The independent variable is plotted along the **horizontal axis** and the dependent variable on the **vertical** axis.

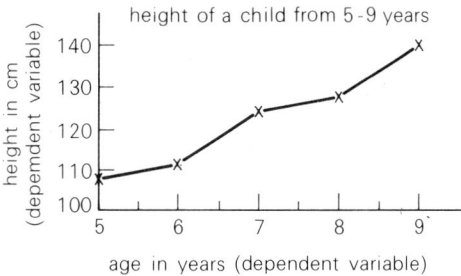

height of a child from 5–9 years

DEPRECIATION

A lowering in the **value** of goods.
Example: A car is bought for £1500. After one year the value is £1200. There has been a depreciation of £300.

DEPRESSION, angle of

The **angle** between a **horizontal line** (CD) and the line (CA) from C to a **point** lower than C.

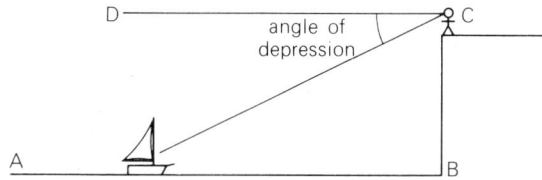

Hence angle DCA is the angle of depression of A from C. (See ELEVATION.)
LATIN *de*, down; *premere*, to press.

DESCARTES, 1596–1650

René Descartes was a famous French philosopher, scientist and mathematician. He is best known for his invention of **Cartesian coordinates** which link **algebra** and **geometry**.

DESCENDING ORDER

Decreasing.
The **length** 24 cm, 16 cm, 15 cm, 7 cm, 1 cm are in descending order. Each **term** is **less than** the one before it.

DESK CALCULATOR

There are two main forms:
1 The hand calculator which is operated by turning a handle.

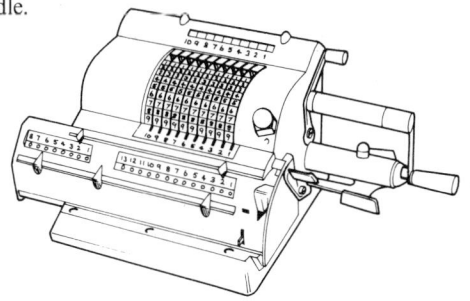

2 The electronic calculator which is operated by pressing buttons. This type may be run from batteries or from the mains' electricity.

DIAGONAL

A **line** joining any two **vertices** of a **polygon** that are not adjacent. *Examples:*

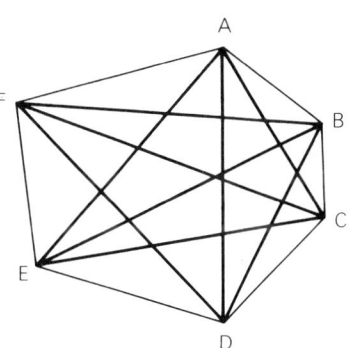

All the lines inside the polygon ABCDEF are diagonals.

A diagonal of a **solid shape** is a line joining any two vertices that do not lie in the same **plane**.

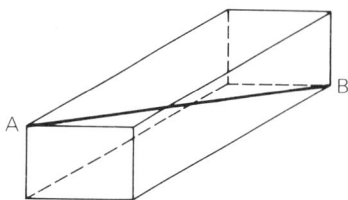

AB is one of the **cuboid's** four diagonals.
GREEK *dia*, through; *gonia*, **corner**.

DIAGRAM

A **term** for any picture, usually a geometric **figure**. Also used for **representations** that are not geometric such as **Venn diagrams**.
Examples:

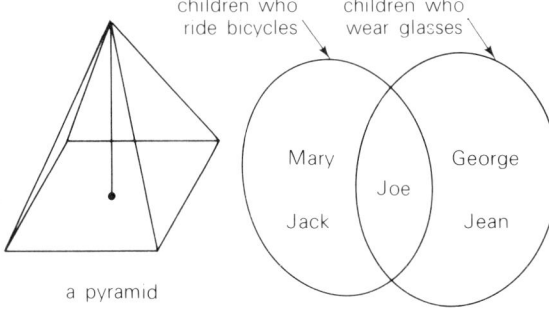

a pyramid

a Venn diagram

DIAL

A graduated **surface**, usually circular, over which a pointer moves to **record** a **measurement**.
Examples: A sundial; **clock** face; speedometer **dial**; dials on electric or gas **meters**.
LATIN *dialis*, daily.
This arose because the sundial was used to tell the **time** of day.

DIAMETER

A **straight line** drawn across a **figure**, particularly when that line passes through the **centre** of the figure. The **term** is mainly applied to **circles** and **spheres**.
1 Diameter of a circle. A line that passes through the centre of a circle and has its end **points** on the circle. (It is therefore a special case of a **chord**.)

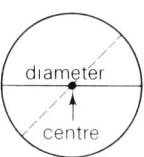

The centre divides the diameter into two **radii**.

2 Diameter of a sphere. A line through the centre of a sphere that ends at the points where it **intersects** the **surface** of the sphere.
GREEK *diametros*, *dia*, through or across; *metron*, a **measure**.

DIAMOND

A **parallelogram** with two adjacent **sides equal**. More correctly called a **rhombus**.
(It is enough to know that two adjacent sides are equal as from this we can **deduce** that all four sides are equal.) A diamond is most easily recognised in **position** (i) and a rhombus in position (ii) yet the two **shapes** are **identical**.

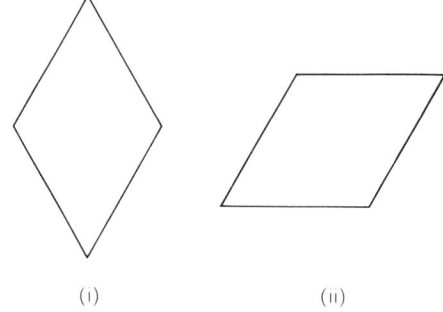

(i)   (ii)

DIE

Plural, dice.
1 A **cube** with **symbols** on its **faces**, generally **dots** or **numerals**. Use in games of chance.

2 Some dice have eight, ten, twelve, twenty or any other **number** of faces according to the purpose they are designed for.

DIENES

Professor Z. Dienes, developer of the **multi-base arithmetic blocks** (M.A.B.), **logic blocks** and other materials used in the teaching of **mathematics**.

DIFFERENCE

1 The **amount** by which one **quantity** is **greater** or **less than** another. It is found by subtracting the smaller from the larger quantity.
Example: The difference of 4 and 9 is 5. (Note that 4–9 is *not* 5.)

2 A way in which two things are not the same.
Example: A difference in colour.
(See SYMMETRIC DIFFERENCE.)

DIGIT

Any one of the basic counting **symbols** of a **number** system. In the system generally used in this country (the **denary system**) the digits are 0, 1, 2, 3, 4, 5, 6, 7, 8 and 9.
In the **binary** system (**base** two) the digits are 0 and 1. Our digits originated from ones used in India well over 1000 years ago.
You will see that some have changed over the years but the symbols for 1, 2, 3, 7 and 0 are much like the ones we now use.

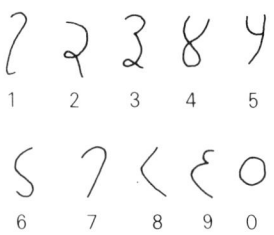

2 The width of the forefinger.
LATIN *digitus*, a finger or toe.

DIGITAL COMPUTER

A device for carrying out mathematical **operations** by means of **digits**.
Example: The **instrument** for measuring the distance travelled in a car or on a bicycle.
The **number** of turns of the wheels are counted and then shown as a distance.

This reads as 1872.4 **kilometres**.
Computers are not always large or expensive pieces of machinery.
Compare with ANALOGUE COMPUTER.

DILATATION

1 Expansion or contraction of a geometric **figure**.

2 A **transformation** that produces a **similar** figure to the original one with **corresponding sides** still **parallel**.
Examples:

These **rectangles** are similar and illustrate dilatation.

These rectangles are similar but do *not* illustrate dilatation since the corresponding sides are not parallel.

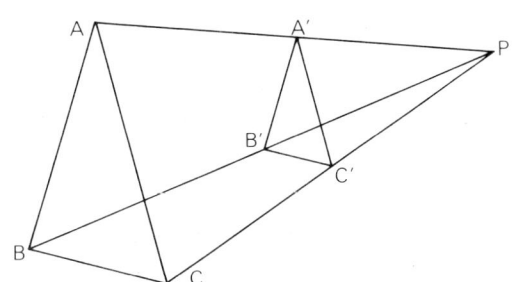

The **angles** of **triangle** ABC are **equal** to the corresponding angles of triangle A′B′C′.

$\angle A = \angle A'$, $\angle B = \angle B'$, $\angle C = \angle C'$.

The corresponding sides are in **proportion**

$$\frac{AB}{A'B'} = \frac{BC}{B'C'} = \frac{AC}{A'C'}$$

Hence triangles ABC and A′B′C′ are similar. Lines joining corresponding **points** (AA′, BB′, CC′) meet at a point (P). Dilation and **enlargement** are two other **terms** that are sometimes used instead of dilatation.

DIME

A coin worth one tenth of a **dollar** (U.S.A.)

1 dime = 10 **cents**.   10 dimes = 1 dollar.

LATIN *decima*, a tenth part. This led to the old French word *disme* which in turn gave rise to dime as a tenth of a dollar.

DIMENSION

1 A **property** related to **length**, **area** and **volume**.
A **point** has no dimensions.
A **curve** has one dimension, (length).
A **plane figure** (**square**, **circle**, **triangle**, . . .) has two dimensions (length and **breadth**).
A **solid** has three dimensions (length, breadth and **height**).

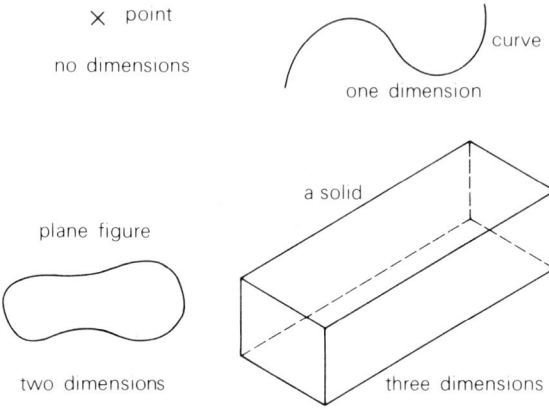

2 The highest **power** is an algebraic expression.
(See DEGREE.)
Example: $y^3 + 2y + 8$ is of the third dimension, $2y^4 + 5y^3 + 9y^2 + y$ is of the fourth dimension. (Note the constant 2 does not affect the dimension.) $3x^2y^3 + 2y^4 + y + 9$ is of the fifth dimension. (Adding the powers of x^2 and y^3 we get $2 + 3 = 5$.) The fourth dimension is **time**. This is a difficult idea that links time and **space**.

DIRECTED LINE

1 A **line** which has **direction**. This may be indicated by an arrow.

If one direction is taken as **positive** then the opposite direction is **negative**.
Example: The **axes** used for **Cartesian coordinates** are given these directions.

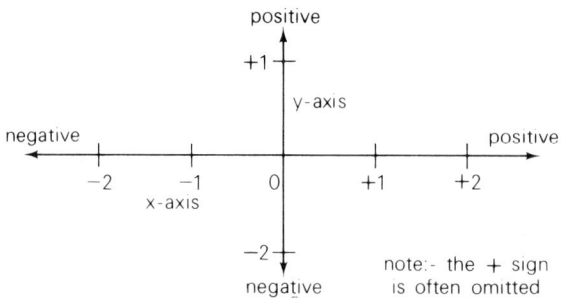

2 A directed line **segment** is part of a line. It has a starting point, end point and a direction. This direction may be shown as $\overrightarrow{AB}$, $\overrightarrow{AB}$, AB or **AB**. All of these show the direction is from A to B.
The direction is taken to be the order in which the letters are written, that is from A to B unless indicated otherwise. $\overleftarrow{AB}$ would show it is from B to A.
(See LINE for detail of origin.)

DIRECTED NUMBER

A **point** on a **line** is labelled **zero**, 0. Equally spaced points to the right are marked $^+1$, $^+2$, $^+3$, . . . (read as **positive** one, positive two, etc.). Similarly points to the left are marked $^-1$, $^-2$, $^-3$, . . . (read as **negative** one, negative two, etc.). $^+1$, $^+2$, $^+3$, . . . , $^-1$, $^-2$, $^-3$, . . . are called directed or **signed numbers**.

$$\begin{array}{ccccccccccc} -5 & -4 & -3 & -2 & -1 & 0 & +1 & +2 & +3 & +4 & +5 \end{array}$$

DIRECTION

1 The **region** or **point** towards which an object is facing or moving.

2 The **position** of one point in **space** relative to another.
Examples:
a A plane flying from London to Edinburgh travels in a northerly direction.
b In what direction is Preston from Hull? West.

DIRECT PROPORTION

(Also called **variation** or **simple proportion**.)
A **relation** between two **variables** in which the **ratio** remains **constant**.

Examples: A car travels at a steady **speed** of 50 **kilometres** per hour. The ratio $\dfrac{\textbf{distance}}{\textbf{time}}$ is constant.

$$\frac{50\,\text{km}}{1\,\text{h}} = \frac{100\,\text{km}}{2\,\text{h}} = \frac{150\,\text{km}}{3\,\text{h}} = \frac{25\,\text{km}}{\frac{1}{2}\,\text{h}} \text{ etc.}$$

If the variables are y and x and the constant is k then $\dfrac{y}{x} = \text{k}$. This can also be written as $y = \text{k}x$. In the example y is the distance, x the time and k the constant speed.

A boy pays 2p for travelling each kilometre on a bus.

The ratio $\dfrac{\text{Cost in pence}}{\text{Distance in km}}$ is constant.

The cost and distance are in direct proportion. This would be written as: cost in pence = **number** of km × 2p.

Hence 20 km costs 40p; 7 km costs 14p and so on. Unless stated otherwise proportion is always taken to mean direct proportion and not **inverse proportion**.

DISC

Also spelt disk.

1 A thin, flat circular plate.

A disc is a **cylinder** with a small **height** compared to the **radius** (or **diameter**).

The height is generally called thickness in such cases.

Examples: coins, gramophone records.

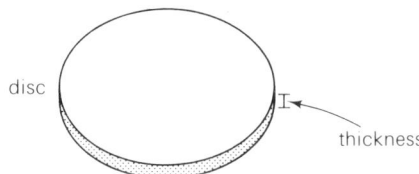

disc — thickness

2 The word **circle** is used in two different ways. (See CIRCLE for these.)

To avoid the confusion of terms many mathematicians now use **disc** for the **region** inside the circle. The term circle then refers to the **line** bounding the disc.

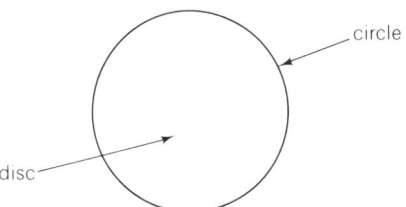

circle

disc

GREEK *diskos*, quoit – as thrown in Greek sports events. LATIN *discus*, quoit, disk.

DISCOUNT

A **reduction** in price made for prompt payment or for buying in large **quantity**.

The discount is usually stated as a **percentage**.

Example: A record player is marked at £20 but there is a sale reduction of 10 per cent.

10 per cent of £20 = $£\dfrac{10}{100} \times 20 = £2$.

The discount is £2 so the actual cost is £18.

DISCRETE

Separate from one another. Discrete objects can therefore be counted.

Examples: The children in a class. The pages in a book. The letters in the alphabet.

Water and other liquids are *not* discrete. They are called **continuous quantities**.

DISJOINT SETS

Two or more **sets** such that no two have a **member** or **element** in common.

Examples: Surnames beginning with A, B and C. **Odd numbers, even numbers.** Boys, girls. Disjoint sets can be shown on a **Venn diagram** as below.

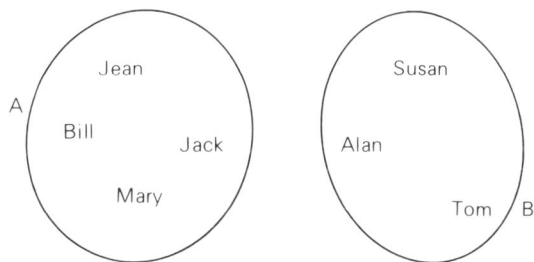

A = {Children aged 12 years or more} = {Jean, Bill, Mary, Jack}.

B = {Children not yet 12 years old} = {Susan, Alan, Tom}.

DISPLACEMENT

1 A movement along a **line** or **curve**.

2 Angular displacement. The **amount** of **rotation**.

3 Displacement jar. A jar used to find the **volume** of an object that does not melt, absorb water or **float**.

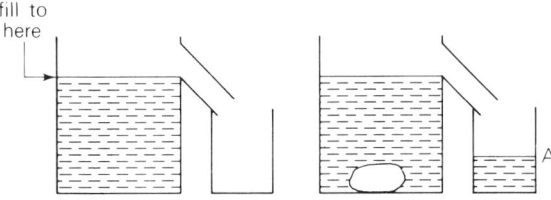

fill to here

The jar is filled to the spout with water. An object is then immersed in the water so that some water is displaced and collected in A. The volume of water in A is then measured and this is **equal** to the volume of the immersed object.

DISTRIBUTION

The way a **set** of **values** are arranged.

Examples: The distribution of ages is shown by this **frequency table**:

Age last birthday	Frequency	
14, 15, 16	111	3
11, 12, 13	⊞	5
8, 9, 10	⊞11	7
5, 6, 7	111	3

Note ⊞ This is 5. A **line** (or **bar**) is drawn through the four to represent the fifth entry.

A distribution can be shown graphically:

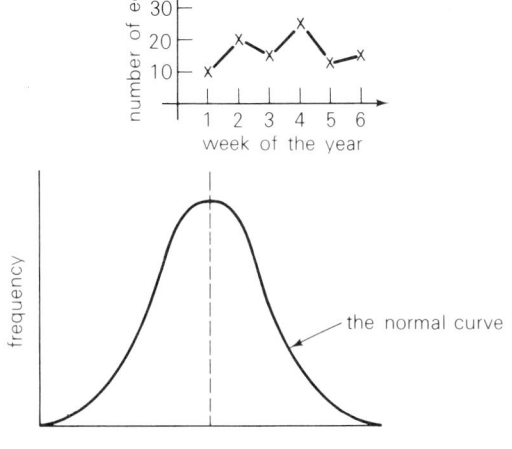

heights of children in British schools

The **normal curve of distribution** is very important in **statistics**. It often occurs when the number of items is large.

DISTRIBUTIVE

Distributive Law or Property. This states a particular **relation** between two **operations** and it is best understood by means of several examples.

a $3 \times (20 + 4) = (3 \times 20) + (3 \times 4)$. This illustrates that **multiplication** is distributive over **addition**. We use this when multiplying

24	4 is multiplied by 3
$\times$ 3	20 (or 2 tens) is multiplied by 3
72	

b Does multiplication distribute over **subtraction**?
$2 \times (13 - 5) = (2 \times 13) - (2 \times 5) = 26 - 10 = 16$.
Since $2 \times (13 - 5) = 2 \times 8 = 16$. This is correct.
It is true for any **number** and therefore multiplication *is* distributive over subtraction.

c Is **division** distributive over addition? If the Distributive Law is true then $24 \div (3 + 2) = (24 \div 3) + (24 \div 2) = 8 + 12 = 20$. But $24 \div (3 + 2) = 24 \div 5 = 4\frac{4}{5}$. Since the two answers are different division does *not* distribute over addition.

In **algebra** we could show the distribution law by $a * (b . c) = (a * b) . (a * c)$. a, b, c are **elements** or **members** of some **set** and * and . are two **operations**.

DIVERGE

To spread out. Opposite to **converge**.

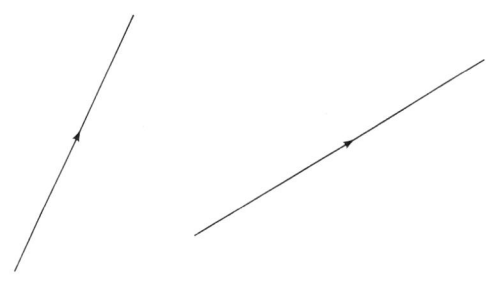

divergent lines

A **series** of **numbers** is said to be divergent if, no matter how large a number you think of, you can get an even larger one by adding enough **terms**.

This series is divergent.
$1 + 2 + 4 + 8 + 16 + 32 + 64 + \ldots$ Each term is obtained by doubling the previous term.

$16 + 8 + 4 + 2 + 1 + \frac{1}{2} + \frac{1}{4} + \frac{1}{8} + \ldots$

Each term is obtained by halving the previous term. This series is *not* divergent, it is **convergent**. No matter how many terms you add the sum is always less than 32.

DIVIDE

To carry out the **operation** of **division**.

Example: Divide 81 by 3

$$\begin{array}{r} 27 \\ 3\overline{)81} \end{array}$$

The word 'into' should be avoided as its meaning is not clear.

Divide 2 'into' 4 can be taken to mean two different things. For example:

a $\dfrac{4}{2} = 2$

or

b You have two cakes and divide them into four **equal** portions. Here 2 into 4 is taken as an **abbreviation** for 'Divide 2 into 4 equal parts'.

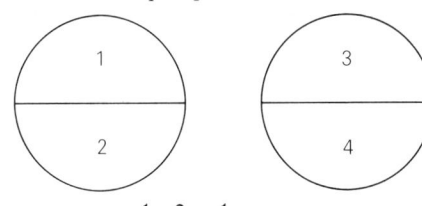

The answer is now $\dfrac{1}{2}$. $\dfrac{2}{4} = \dfrac{1}{2}$.

DIVIDEND

1 The **quantity** that is to be divided.

Example: $48 \div 4 = 12$. 48 is the dividend, 4 is the **divisor**, 12 is the **quotient**.

2 The money paid to the holder of **shares** in return for the money invested.

DIVIDERS

An **instrument** for marking off or transferring **equal lengths**.

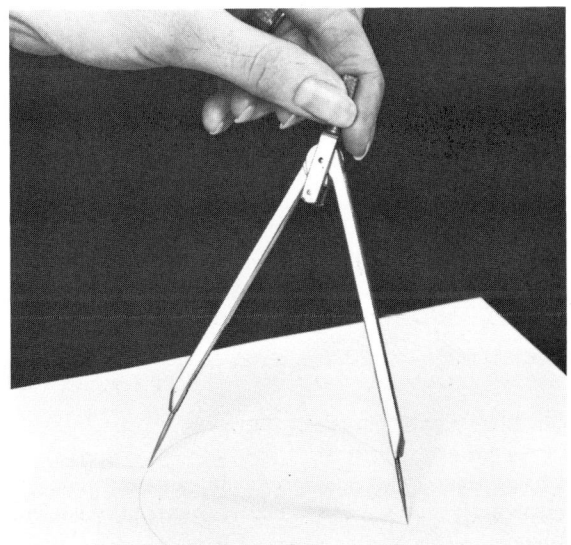

DIVISIBLE

A **number** is divisible by another number if the **remainder** is **zero**.

Examples: 24 is divisible by 6. 35 is *not* divisible by 6.

Some rules for divisibility:

For (2)   Any number ending in an **even digit** is divisible by 2.

(4)   If the last two digits are divisible by 4 then so is the number. (56 is divisible by 4 then so is 187 356.)

(8)   If the last three digits are divisible by 8 then so is the number. (352 is divisible by 8, therefore so is 978 352.)

(3)   If the **sum** of the digits is divisible by 3 then so is the number. (Consider 3879. The sum of the digits is $3 + 8 + 7 + 9 = 27$. This is divisible by 3, therefore so is 3879.

(5)   The number must end in 5 or 0.

(10)   The number must end in 0.

(6)   Test for divisibility by 2 and by 3.

(9)   The sum of the digits must be divisible by 9.

(See CASTING OUT NINES.)

DIVISION

An **operation** on **numbers** that can be interpreted in several ways.

a **Quotition** or **grouping**. The number of **groups** of a stated **size** is found.

Example: $24 \div 3$. The number of 3's in 24 is 8. This is **equivalent** to successive **subtraction** of 3 and finding how many **times** this can be done.

b **Partition** or **sharing**. The number or **quantity** to be divided (called the **dividend**) is shared equally into the stated number of parts.

Example: $24 \div 3$. 24 is shared into three **equal** parts giving $24 \div 3 = 8$.

c **Ratio**. **Comparison** is made between two quantities. Comparing **lengths** of 20 cm and 30 cm we have a ratio of $\dfrac{20}{30}$ or $\dfrac{2}{3}$. This is read as a **ratio** of 2 to 3.

d The **inverse** of **multiplication**.

For example $20 \div 5$ is **equivalent** to finding what 5 must be multiplied by to give 20. This could be written $5 \times \square = 20$ or $5x = 20$.

We then find the value of $\square$ and x that make the **equations** true.

(See DIVISIBLE, LONG DIVISION, SHORT DIVISION.)

DIVISOR

The **number** that is divided into another.

$$36 \div 12 = 3$$

12 is the **divisor**.

36 is the **dividend**.

3 is the **quotient**.

(See DIVIDE, DIVISION.)

DODECAGON

A **polygon** with twelve **sides**.

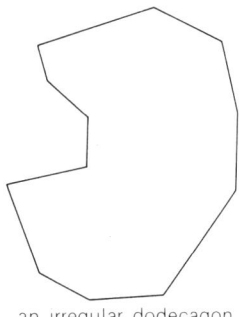 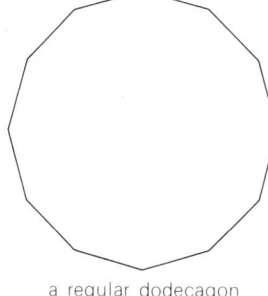

an irregular dodecagon   a regular dodecagon

To draw a regular dodecagon

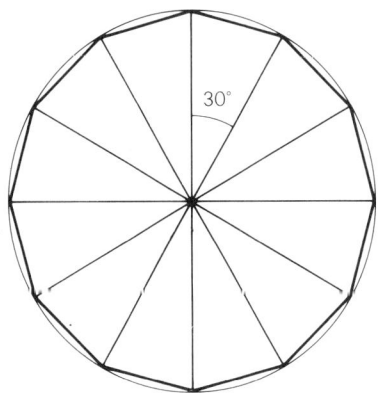

Draw a **circle**. $360° \div 12 = 30°$. Use a **protractor** to mark off **intervals** of 30°. This will give 12 **points**, equally spaced on the circle. Join these to make the dodecagon. (A dodecagon can be constructed using only a **ruler** and **compasses**. Construct **angles** of 60°, **bisect** them and proceed as above.)

GREEK *dodeka*, twelve; *gonia*, an angle.

DODECAHEDRON

A **solid** with twelve **plane faces**. When the faces are all **equal regular pentagons** the solid is a regular dodecahedron.

The regular dodecahedron is one of the five possible regular solids (See PLATONIC SOLIDS.)

GREEK *dodeka*, twelve; *hedra*, a seat.

DOLLAR

A **unit** of money in the United States of America, Canada, Australia and many other countries.

In the U.S.A. 1 dollar = 100 **cents** = 10 **dimes**.

A German coin, the Joachim's-thaler was so named because it was made from silver mined in Joachimstal, Bohemia. This coin became known as a thaler (also taler) and later it was called daler. Further changes led to dollar.

DOMAIN

The **set** of possible **values** for one **variable**. The form is generally applied to a set with **elements** x that are plotted on the **horizontal axis** of a **graph**. Those plotted on the **vertical** axis are called the **range**. These elements are denoted by y.

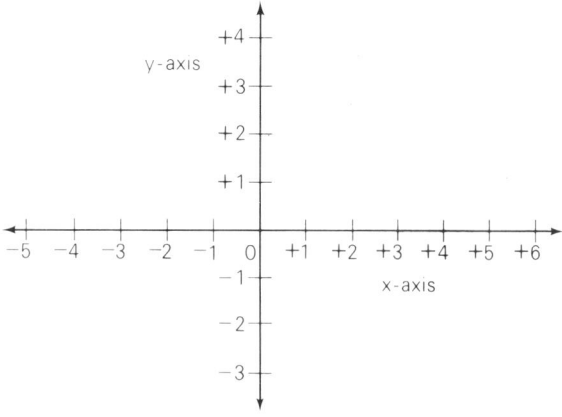

(See DEPENDENT and INDEPENDENT.)

LATIN *dominus*, master.

(This is because the values in the domain determine the values for the range, just as a master determines what will follow.)

DOMINO

1 Two **adjacent squares**. In games of chance each square has a **numeral** or dots written in it to show its value. There are 28 such pieces in the game of dominoes with the number of dots indicating 0, 1, 2, 3, 4, 5 or 6.

2 A domino is a special case of a **polyomino**.

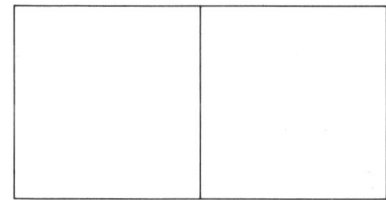

a domino,
there is only
one

DOT

1 A small spot or mark used to indicate a **point**.

·A

Mathematicians think of a point as having no **dimensions** but when we make a dot it always has some area or we could not see it. A point is a mathematical idea and we cannot actually draw one.

2 Dot and carry. When adding a dot may be used to show each **group** of ten reached.

27	$1 + 7 = 8$	(c) Adding from
48·	$3 + 8 = 11$. $11 = 10 + 1$	(b) the bottom
54·	$9 + 4 = 13$. $13 = 10 + 3$	(a) upwards.
69		
198	The **units** therefore come to 28.	

Dots could also be used when adding the tens **column**.

3 In advanced work a dot product is one way of multiplying **vectors**.

DOUBLE

Twice as many.
Examples: 8 is double 4. Double 6 and the **result** is 12.
The **term** also shows something occurs twice in many other expressions:
double back, double-headed penny, double glazing, double up laughing, double agent and many others.
LATIN *duplus*, double; from *duo*, two.

DOZEN

Twelve.
With the introduction of the **metric system** many goods are now sold in **sets** of ten instead of twelve. A baker's dozen is thirteen.
OLD FRENCH *dozeine*, twelve.
This in turn probably came from duodecim made up from the two Latin words *duo*, two and *decim*, ten.

DRY MEASURE

A system for measuring dry goods such as fruit and grain. The main **units** are **pints**, **quarts**, **pecks** and **bushels** but they have been largely replaced by those of the **Système Internationale**.

2 pints = 1 quart
8 quarts = 1 peck
4 pecks = 1 bushel.

DUAL

1 Relating to two.

Example: A dual controlled car has two **sets** of controls.

2 In higher **geometry** when two **elements** or **operations** can be interchanged.

Example: Three **points** can be joined to give three **lines**. The dual: Three lines can be made to **intersect** at three points.

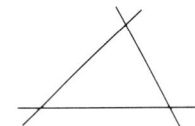

LATIN *dualis, duo*, two.

DUODECIMAL SYSTEM

A **number** system with a **base** of twelve.

Example: 32 in base twelve indicates there are 3 twelves and 2 **units**. 32 (base twelve) is **equivalent** to 38 in base ten.
$[(3 \times 12) + 2 = 36 + 2 = 38]$

In base ten there are ten **symbols** 0, 1, 2, 3, 4, 5, 6, 7, 8 and 9. In base twelve there are twelve symbols. We use the ten above and two others. t and e could be used. (To represent 10 and 11 respectively in base ten).

Example: 1t base twelve is $(1 \times 12) + (10) = 22$ in base ten.
     12 10

LATIN *duo*, two; *decim*, ten.

DUPLATION

Doubling.

Duplation was used by the Egyptians as a means of multiplying without the need to learn the **multiplication tables**.

Example:

$13 \times 7 = (13 \times 2) \times 2 + (13 \times 2) + 13 = 52 + 26 + 13 = 91.$
       ↑   ↑   ↑
    Four 13's   Two 13's   One 13

(See RUSSIAN MULTIPLICATION.)

LATIN *duplus, double*.

DUPLICATE

1 Copy. By using carbon paper we can make a duplicate of whatever we write.

2 **Double.**

LATIN *duplus, double*.

DYNAMICS

The study of **forces** and their **results** on bodies. Often limited to cases when motion takes place.

(See STATICS for when no motion takes place.)

E

E

1 Written as E the **symbol** for energy.
Example: **Einstein's** famous **formula** in the Theory of Relativity is $E = mc^2$.

2 e. A **number** used for the **base** of natural **logarithms**.
e is **approximately** 2.718.
e is also used as a symbol for other **quantities**, such as eccentricity and the coefficient of restitution.

3 Written as $\mathscr{E}$ it is the symbol for the **universal set**. Chosen because it is the first letter of the French word 'ensemble' meaning 'all the parts taken together'.

EDGE

A **line** formed by the **intersection** of two **plane surfaces**.
Example: A **cuboid** has twelve **edges**.

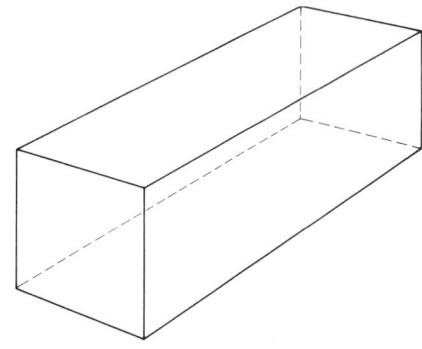

The **term** is loosely used for any intersection of surfaces as in 'the edge of a cliff'.
A **straight-edge** refers to anything that can be used to draw a **straight** line.

EGG-TIMER

Also called an egg-glass. Sand is trapped in a glass tube and it slowly trickles through from the upper to the lower part.
The **time** for this to happen denotes how long an egg should be boiled for.

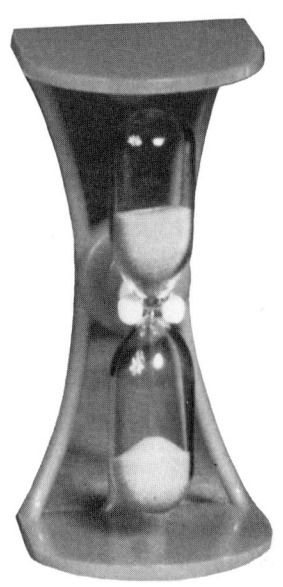

EGYPTIAN SYSTEM

The Egyptians used these **symbols** for **numbers**.

I	II	III	IIII	III	III	IIII	IIII	III
				II	III	III	IIII	III
								III
1	2	3	4	5	6	7	8	9

∩	∩∩	∩∩∩	℮ or ℮
10	20	30	100

1000	10 000

Example: ℮℮℮ ∩∩ III/II was 325

The symbols used are called hieroglyphics. There were also symbols for larger numbers such as a bird for 100 000 and a man for 1 000 000.

EINSTEIN, 1879–1955

Albert Einstein was one of the greatest scientists the world has ever known. His work on **time**, **space**, **mass**, **light** and **gravity** completely changed the way scientists had approached these subjects. He is best known for his **theories** of relativity and his famous energy **equation** $E = mc^2$. This stands for

Energy = mass × **velocity** of light, squared.

$E = m \times c^2$

Einstein.

ELEMENT

1 An item belonging to a **set**.

The **term member** is often used instead of element.

Example: Set $A = \{a, e, i, o, u\} = \{$Vowels in the English alphabet$\}$. a, e, i, o and u are elements (or members) of set A.

2 *The Elements.*

A collection of thirteen books by **Euclid** mainly on **geometry**. With little change these were used in schools for over 2000 **years**.

3 The basic terms, ideas and assumptions of a subject, particularly as developed in **geometry**.

ELEMENTARY

1 In a wide sense used to denote 'simple'. This of course depends upon age and ability. Elementary **geometry** may seem very difficult to an 11 year old and yet the same person may find it simple when 16 years of age.

2 Work founded upon basic **terms**, ideas and assumptions (**elements**).

Elementary **algebra** would be built logically from these elements.

ELEVATION, ANGLE OF ELEVATION

1 The **vertical** distance between one **point** and another, or between a point and a **plane**.

Example: The elevation of a mountain top, measured from sea **level** is 1500 **metres**. From a village 1000 metres above sea level the elevation of the mountain top is only 500 metres.

2 **angle** of elevation.

Angle ABC in the picture is the angle of elevation. BC is **horizontal**. C is directly below the **kite** A.

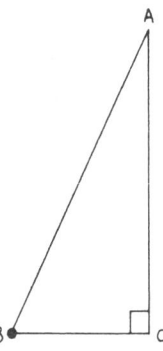

(See DEPRESSION for angle of depression.)

3 The view of **solid** objects from the front.
Example:

elevation of
a house

(See PLAN, the view from above.)
LATIN *elevare*, to raise.

ELLIPSE

A geometric **figure**.

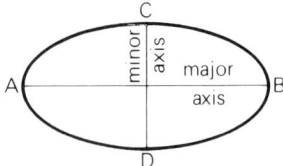

The figure made by a **section** through a **cone**, cut as shown.

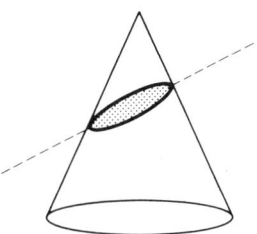

Curves obtained from a section of a cone are called **conic sections**. The ellipse is one of these; others are the **hyperbola**, **circle** and **parabola**.

An ellipse has two **axes** of **symmetry**, the larger one is the **major** axis (AB) and the shorter one the **minor** axis (CD). In advanced work an ellipse can be regarded as the **locus** (or path) of a **point** such that the **sum** of the distances from two fixed points is **constant**. This suggests a way of drawing an ellipse:

Thread is tied to two drawing pins which pin a sheet of paper down firmly.

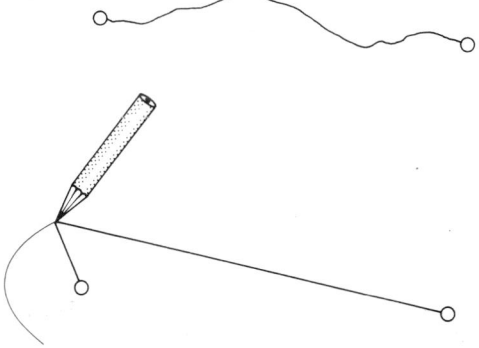

A pencil point is used to tighten the thread and then moved into various **positions**. The point traces out an ellipse.

An ellipse can also be considered as the locus of a point (P) such that its distance from a fixed point (F) is **less than** the distance from a fixed line (AB), the **ratio** of the distances being constant and less than 1.

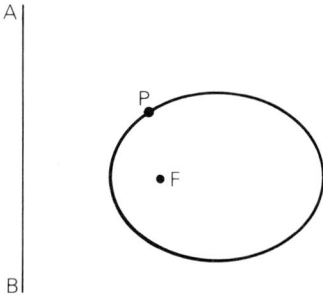

GREEK *elleipsis*, defect, having part left out. Probably originates from the circle being thought of as 'perfect' and the ellipse being defective and not so perfect.

EMPTY SET

A **set** containing no **elements** or **members**. Also called a **null set**. It is written as { } or ∅.
Examples: The set of people with two heads. The set of children more than 40 **years** old.

ENCLOSED

Being on the inside, as when sheep are enclosed by fences.
Examples: Mathematical **terms** are enclosed in **brackets** to show the **order** of **operation**. $12 - 5 \times 2 = 2$ but it is clearer if written $12 - (5 \times 2) = 2$. This avoids the error of thinking that $12 - 5 \times 2$ is 14. (**multiplication** is done before **subtraction**.)

In **Venn diagrams members** of a **set** are enclosed to show they belong together.

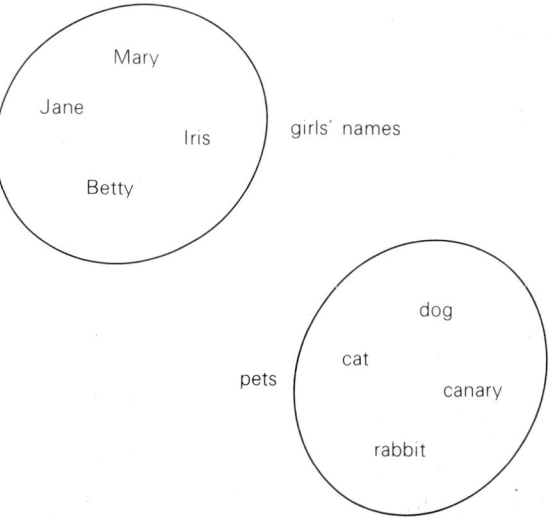

ENLARGEMENT
See DILATATION.

ENUMERATE

1 To **number** or **count**.

2 To list.

Example: Enumerate the vowels in the alphabet. a, e, i, o, u.

ENVELOPE

A **curve**, **line** or **surface** that is a **tangent** to (touches) a **member** of a **set** of other curves, lines or surfaces.
Example:

1 In curve stitching:

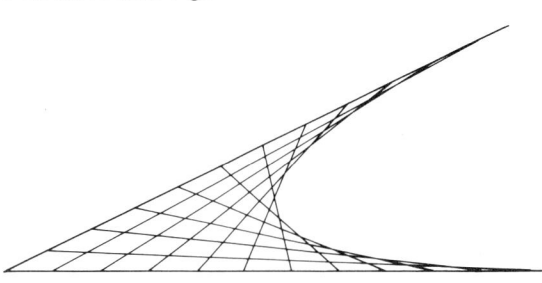

The **parabola** is the envelope of the set of **straight** lines.

2 **Equal chords** are drawn in a **circle**. The inner circle is the envelope of the set of chords.

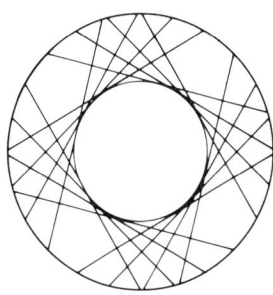

EQUAL

1 Two things are said to be equal if they are the same in some stated manner.

Example: Two men may be equal in **height** but different in many other ways.

2 $5 + 3, 4 + 4$ and $9 - 1$ are equal **numbers** since they are all ways of writing 8. Numbers are said to be equal if they are different ways of writing each other.

3 **Sets** are equal if they have exactly the same **members** or **elements**. The **sign** $=$ is used to denote equality.
Example: Set $A = \{a, e, i, o, u\}$. Set $B = \{u, i, a, o, e\}$.
       Set $C = \{$The vowels in the English alphabet$\}$.
       Set A = Set B = Set C.

EQUAL ADDITION

A method of **subtraction**. It involves adding **equal quantities** to two **amounts**, their **difference** then remains unchanged.
Example:

$$\begin{array}{r} 7\,{}^18 \\ -\,{}^32\ 9 \\ \hline 4\ 9 \end{array}$$ 10 is added to the 8 making 18 and 10 is added to the 2 tens making 3 tens.

In effect we have replaced $78 - 29$ by $88 - 39$. The method of equal addition is thought to be faster than the other main subtraction methods, **decomposition** and **complementary addition**. It is, however, more difficult to understand.

EQUAL SIGN

$=$ is used to denote equal **quantities**, especially **numbers**.
Example: $2 \times 4 = 8$.   100 centimetres $= 1$ **metre**.
First used by Robert Recorde, an English mathematician, in a book on **algebra**, *Whetstone of Witte*, 1557.

EQUATION

A mathematical **statement** asserting that two expressions 'name' or 'stand for' the same **number**. The two expressions are separated by the **equal sign** $=$ which denotes their equality.
Example: $2x = 8$. This equation is true only if x is 4.
    $\square + 7 = 19$. $\square = 12$.
The $\square$ and x are called **place holders**.

EQUATOR

The **great circle** which is an **equal** distance between the North and South Poles.
The equator is **perpendicular** to the **axis** of **rotation** of the Earth. The **latitude** of all points on the equator is $0°$ (**zero degrees**).

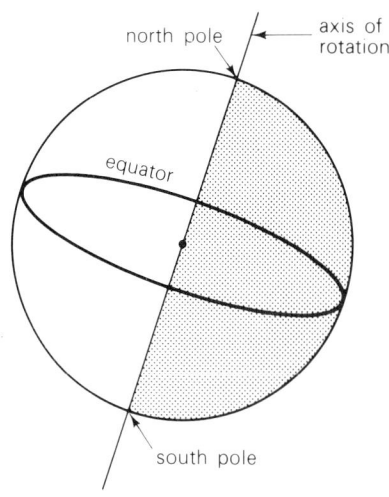

So called because day and night are of equal **length** when the sun is over the equator.

EQUIANGULAR

Having **equal angles**. Particularly applied to **polygons**. All the angles of the **regular octagon** are 135°. It is therefore an equiangular polygon.

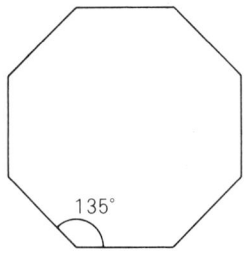

EQUIDISTANT

Equal distance.

Example: All **points** on the **circumference** of a **circle** are equidistant from the **centre** (C).

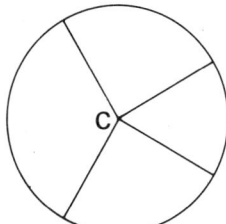

EQUILATERAL

Equal sides.

Examples:

1 An equilateral **triangle**.

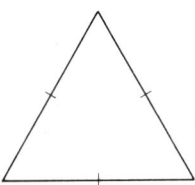

The three **angles** are also equal so an equilateral triangle is also **equiangular**.

2 An equilateral **pentagon**.

In A the five sides are equal in **length** but the angles are not equal. A is equilateral but not **equiangular**.

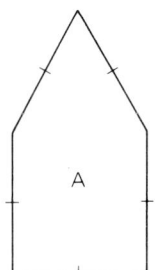 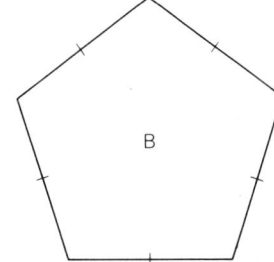

In B the angles are all equal so this pentagon is equilateral and equiangular.

EQUILIBRIUM

The state of a body when the **forces** acting on it **balance** so that it stays at rest.

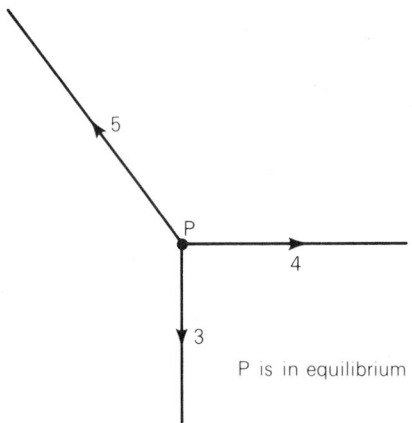

P is in equilibrium

LATIN *aequalis*, equal; *libra*, balance.

EQUIVALENCE RELATION

A **relation** which establishes that the **elements** or **members** of a **set** are the same in a stated respect. For example the same sex, all made of wood or the same **height**.

Three conditions have to be satisfied (a) **reflexive** (b) **symmetric** and (c) **transitive**.

Example: 'is the same colour as'.

Suppose A, B and C are all yellow

A is the same colour as A. B is the same colour as B. C is the same colour as C (Reflexive).

A is the same colour as B then B is the same colour as A. This must apply to any **pair** (Symmetric).

A is the same colour as B and B is the same colour as C implies A is the same colour as C (Transitive). Hence 'is the same colour as' is an equivalence relation.

EQUIVALENT

Having the same **value**. Resulting in the same **set**.

Example:

a 7×4, $7+7+7+7$, 14×2 and 2×14 are equivalent as they have the same value, 28.

b In a class the set of boys is equivalent to the set of children who are 'not girls'. Two sets are equivalent if they have the same **members**.

c $\frac{2}{3}$, $\frac{4}{6}$, $\frac{6}{9}$, $\frac{8}{12}$ and $\frac{10}{15}$ all stand for the same **fraction**, generally represented by $\frac{2}{3}$ as this is the simplest form. Such fractions are called equivalent **fractions**.

ERATOSTHENES, 275–195 BC

Greek scholar who wrote on many subjects besides **mathematics**. Well known for his **measurement** of the Earth's **circumference** and for the method of finding **prime numbers**, known as the **Sieve of Eratosthenes**.

ERROR OF MEASUREMENT

The **difference** between the measured **amount** and the true amount.

Example: A rod known to be 38.5 **cm** is measured as 38.9 cm.

The error of measurement is 0.4 cm.

The **percentage** error is

$$\frac{(\text{Measured amount} - \text{true amount})}{\text{true amount}} \times 100$$

that is $\dfrac{\text{error}}{\text{true amount}} \times 100$.

ESTIMATE

A judgement of an **amount** without **measuring**.

Example: The **length** of a table may be estimated by comparing it with other lengths you know, or think how many **times** a known length (such as a pencil) will have to be repeated to **equal** the length of the table.

A **guess** differs from an estimate. When guessing you have no idea of the answer and do not use judgement or previous knowledge. Both of these are used in estimating.

Euclid

EUCLID

Euclid of Alexandria was probably born about 365 BC but the exact dates of his birth and death are not known. He may have been an Egyptian but it is more likely that he was a Greek who went to Alexandria in Egypt. This great centre of learning was a Greek colony at that time.

Euclid is the most successful writer of text books that the world has ever known. Over a thousand editions of his **geometry** books have been printed in the last 500 **years**. His greatest work is called *The Elements.*

LEONARD EULER.
London. Published as the Act directs, Oct.ᵗ 23ᵗʰ 1804. by J. Wilson.

EULER, 1707–1783

Léonard Euler was a famous Swiss mathematician who studied many other subjects including theology, medicine, music, oriental languages, astronomy and physics. He founded the branch of **mathematics** called **topology** and gave his name to **Euler's formula** which connects the **number** of **vertices**, **faces** and **edges** of a **solid**. Due to excessive work Euler lost the use of his **right** eye in 1735. This loss can be seen in his picture. In 1766 he lost the use of the other eye but continued to work without complaint until the day of his death.

EULER'S FORMULA

This **formula** was made famous by the Swiss mathematician **Euler** but it was known earlier to both **Archimedes** and **Descartes**. The formula, $V+F-E = 2$ connects the **number** of **vertices** (V), **faces** (F) and **edges** (E) for any **solid**. (Modifications of this cover **two-dimensional figures**.)
Examples:

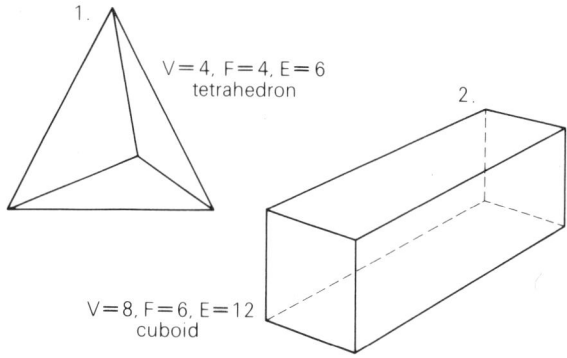

1. $V = 4$, $F = 4$, $E = 6$
tetrahedron

2. $V = 8$, $F = 6$, $E = 12$
cuboid

In both cases $V+F-E = 2$.

LATIN *forma*, form.

EVALUATE

To find the **value** of, especially applied to simplifying a **numerical** expression.

Examples: Evaluate 37×4.

$$\begin{array}{r} 37 \\ \times\ 4 \\ \hline 148 \end{array}$$

Evaluate $3F+2D-4C$ when $F = 5$, $D = \frac{1}{2}$ and $C = 1$.
$(3 \times 5)+(2 \times \frac{1}{2})-(4 \times 1) = 15+1-4 = 16-4 = 12$.

EVEN NUMBER

A **number** that has no **remainder** when **divided** by 2. For example 6, 14 or 24. A **whole number** which is not **odd**. If the whole numbers are written down in **sequence** the odd and even numbers **alternate**.

0	1	2	3	4	5	6	7	. . . etc.
even	odd	even	odd	even	odd	even	odd	

Notice that as $0 \div 2$ leaves no remainder 0 is even.

EXACT

Without **error**, **accurate**.
One **number** can be divided exactly by another if there is no **remainder**.
Example: 24 can be divided exactly by 6.

EXCHANGE

To give one thing in return for another that is **equivalent** in **value** or **amount**. A 10p coin can be exchanged for ten 1p coins. When shopping we **exchange** money for goods of the same value.

EXPANDED NOTATION

The normal way of writing a **number** is replaced by an expanded 'larger' one.
Example: $547 = (5 \times 100)+(4 \times 10)+7$
or $(5 \times 10^2)+(4 \times 10)+7$

EXPENDITURE

The money spent.

EXPONENT

Also called **index**. When 1000 is written as 10^3 the 3 is the index, or exponent. It shows the **power** to which 10 is raised.
Example: 7^4 has an exponent of 4. It indicates four 7's are multiplied together, $7^4 = 7 \times 7 \times 7 \times 7$.
a^6 has an exponent of 6.
$a^6 = a \times a \times a \times a \times a \times a$ (that is six a's multiplied together).

EXTERIOR

Outside.

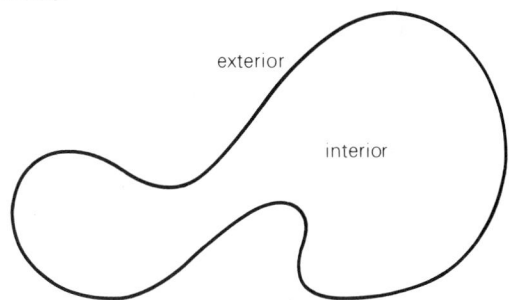

exterior

interior

EXTERIOR ANGLE

The **angle** formed outside a **polygon** by two **adjacent sides**, one of which has been **produced**.

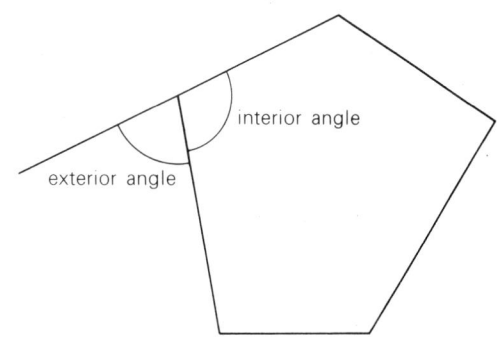

interior angle

exterior angle

The **sum** of the exterior angle and its **interior** angle is $180°$.

EXTRACT A ROOT

The **square root** of 25 is 5 because $5 \times 5 = 25$. The square root of 3 is **approximately** 1.732. When we calculate the square root of a **number** we extract the **root**. Similarly other roots can be extracted. For example the **cube root** of 64 is 4 since $4 \times 4 \times 4 = 64$.

EXTRAPOLATE

To extend and find a **value** outside those given.
Example:
1

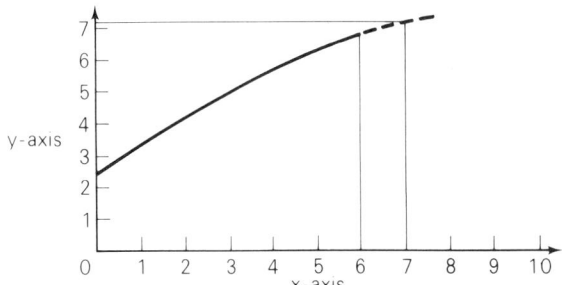

The **curve** is given up to the value $x = 6$. By extending it (shown as –––––) the value of y when $x = 7$ can be **estimated**. It is **approximately** 7.2.

2

x	0	1	2	3	4	5	6	7
y	5	9	13	18	24	30	37	44

By extrapolation we can estimate a value for y when $x = 8$.

Subtracting each y value from the one following it we see the y increases are 4, 4, 5, 6, 6, 7, 7 so the most likely one after that is 8. Therefore y = 44 + 8 = 52 when x = 8.

We cannot be certain of the value when we extrapolate.
(See INTERPOLATION.)

EXTREMES

1 The end **points** of a **line segment**. A and B are the extremes.

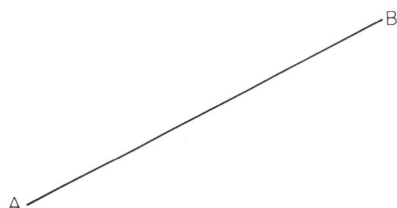

2 In **proportion** the **ratios** $\frac{3}{5} = \frac{9}{15}$ can be written 3:5 = 9:15. The 3 and 15 are called the extremes.

3 Sometimes used to denote a **maximum** or **minimum** **value**.

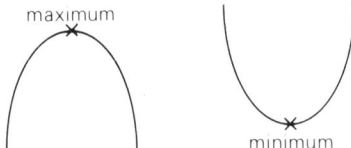

EYE-LEVEL

On a **horizontal plane** containing the eye.

F

F
F **Abbreviation** for **Fahrenheit**. Freezing point is 32°F.
f Used to indicate a **function**.

FACE
A **side** of a **solid**. Usually restricted to flat **surfaces** (**planes**) such as those of a **cube**, **pyramid** or any **polyhedron**.
Example: A cube has six faces.

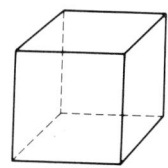

This pyramid has five faces.

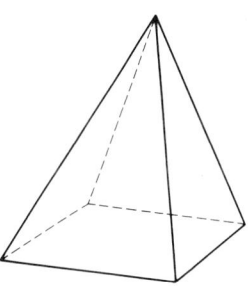

FACTOR
When a **number** is expressed as the **product** of two or more other numbers then these numbers are called the factors of the first number.
Example: 32 = 4 × 8. 4 and 8 are factors of 32. (Other factors are 1, 2, 16 and 32). It follows that a factor will divide exactly (without **remainder**) into a number it is a factor of.
Example: 148 ÷ 4 = 37. 4 is therefore a factor of 148. (See COMMON FACTOR, HIGHEST COMMON FACTOR.)

FACTORIAL
The factorial of a **natural number** (1, 2, 3, 4, . . .) is the **product** of that **number** and all the natural numbers less than it. It is denoted by ! after the number.

Example: Factorial 4 is written as 4! or else as $\underline{|4}$
$4! = 4 \times 3 \times 2 \times 1 = 24$.
For a natural number n
$$n! = n(n-1)(n-2)(n-3) \ldots \times 4 \times 3 \times 2 \times 1$$
Factorials are used in work on **permutations** and **combinations**.

FACTORISE
Also spelt **factorize**.
To express as a **product** of **factors**.
32 can be factorised in many ways, such as 8×4, 2×16. 4, 8, 2 and 16 are called factors of 32. In **algebra**
$$3x(2x+5) = 6x^2 + 15x$$
$6x^2 + 15x$ can be factorised, its factors being $3x$ and $2x+5$. We frequently need to factorise **numbers** into **prime factors**.
Example: $144 = 2 \times 2 \times 2 \times 2 \times 3 \times 3$ or $2^4 \times 3^2$.
2 and 3 are **prime numbers**.

FAHRENHEIT
A **scale** for measuring **temperature** now little used. It has been replaced by the **Celsius** scale. The Fahrenheit scale took 32° as the freezing point of water and 212° as the boiling point of water. 180° Fahrenheit are **equivalent** to 100° Celsius.
To **convert** Fahrenheit (F) into Celsius (C) use this **formula**.
$C = \frac{5}{9}(F-32)$.
Example: Convert 68°F to Celsius.
$$C = \frac{5}{9}(68-32) = \frac{5}{9} \times 36 = 20.$$
$$68°F = 20°C.$$
Named after the inventor of the scale, D. G. Fahrenheit (1686–1736).

FALSE SENTENCE
A mathematical **statement** that is not true.
Example: $7 < 4$ is a **false sentence** (or statement).
$x+4 = 6$ is called an **open sentence**.
It is false, if, for example, x is replaced by 8.
It is a **true sentence** if x is replaced by 2.

FAMILY

A **set** of **lines** or **curves** that have a feature in common.
Example: This family of lines all pass through the same **point**.

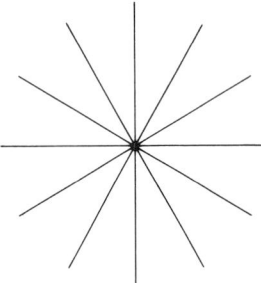

This family of curves all **touch** the same line.

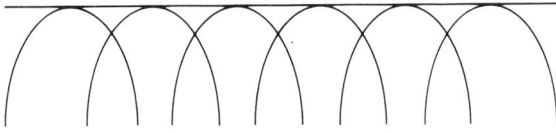

FAST

Quick. Swift. Covering a distance in a short period of **time**. Fast is a **relative term**. A **slow** dog may run at a greater **speed** than a fast mouse.

FATHOM

A nautical **measurement** of **length equal** to 6 feet or approximately 1.83 **metres**. Sailors dropped a rope, weighted by a stone, into water to find the depth. As they wound out the rope they counted each part between their stretched arms as 1 fathom. Now a fathom is exactly 6 feet.

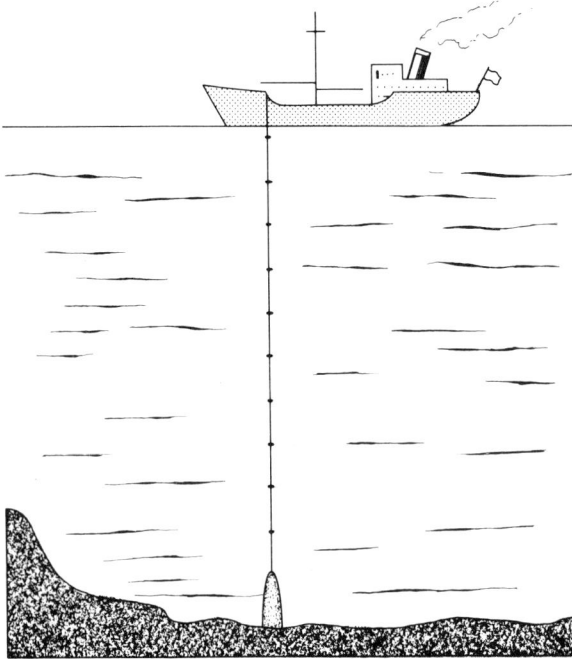

FIBONACCI SEQUENCE

The **sequence** of **numbers** 1, 1, 2, 3, 5, 8, 13, 21, 34, 55, . . . (Sometimes given as 0, 1, 1, 2, 3, 5, 8 etc.) Starting with 2 each number is the **sum** of the previous two numbers.
The numbers form a **pattern** that occurs frequently in nature. The sequence is named after the Italian mathematician, Fibonacci who was also known as Leonardo of Pisa, (1170–1230).

FIELD

1 A system in advanced **mathematics** in which **elements** and **operations** satisfy certain conditions.

2 Sometimes used instead of **domain**. It is best avoided so that there is no confusion with the first definition, 1.
OLD ENGLISH *feld*, field.

FIFTEEN PUZZLE

A puzzle that was very popular about a **century** ago. It consists of a **square** box into which fifteen blocks numbered 1 to 15 can be placed. The bottom right square is left empty. The **problem** is to rearrange the blocks so as to get the **position** shown in the **diagram**.

The blocks must not be lifted; the only move allowed is to **slide** a block into the empty square. For example in the bottom diagram only 15 or 12 can be moved.

FIGURATE NUMBERS
See POLYGONAL NUMBERS.

FIGURE
1 A **set** of **points** in **geometry** such as those forming a **line**, **curve**, **square** or any other **shape**. In **three dimensions** the set may form a **cuboid**, **sphere** or other shape.

2 A **symbol** used to denote an **integer**. $(7, 2, 8, \ldots)$.

FINAL
Last. At the end.

FINGER CALCULATING
An ancient way of multiplying. Two **numbers greater than** 5 are multiplied this way: Suppose we need to calculate 7×8. **Subtract** 5 from each number giving 2 and 3. With the palms facing outwards put up two fingers on the left hand and three on the **right** as shown.

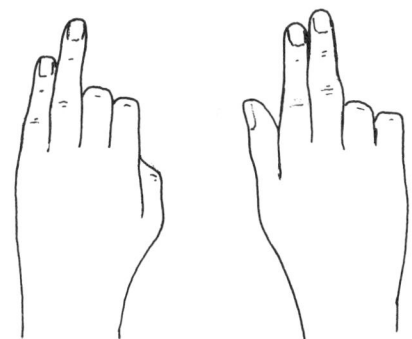

(Note the fingers on the left of each hand are used.)
The **total** number of fingers held up shows how many tens there are in the **product** $(2+3 = 5$ so there are 5 tens). **Multiply** the number of fingers not raised $(3 \times 2 = 6)$. This gives the number of **units**.
Therefore $7 \times 8 = 56$.

FINITE
1 Having bounds or **limits**. The **region** inside a **circle** is finite since it is bounded by the **circumference**. Not **infinite**.

2 A finite **set** is a set with a countable **number** of **elements** or **members**.
Examples: $\{5, 8, 13, 27, 31\}$.
The letters of our alphabet.
The set of **odd** numbers is *not* a finite set.
They cannot be counted. $\{1, 3, 5, 7, 9, 11, 13, \ldots\}$.

FIRST
At the start. Before any other.

FLAT
1 Smooth, **level**. A **surface** is flat if, when any two **points** on it are joined, the joining **line** lies entirely in the surface.

2 Flat **angle**. $180°$. Two **right angles**.

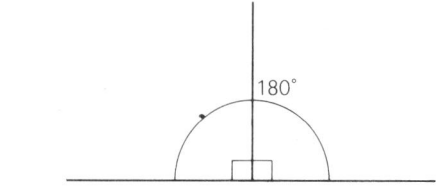

3 The name of a particular part of the **multibase arithmetic blocks**. The example shows a flat in **base** four. Similarly other bases also have flats.

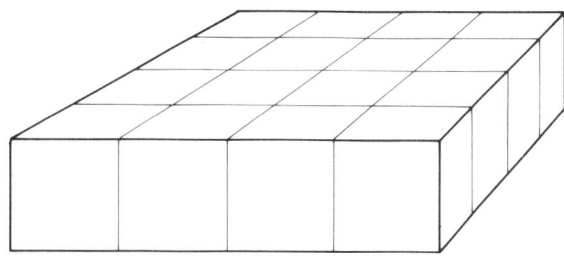

FLOAT
To be suspended on the **surface** in a liquid. Not sinking.
Example: A stone sinks in water, a cork floats.

FLOW CHART
A chart or **diagram** showing the **order** of steps to be taken.
Example: How to eat a toffee.

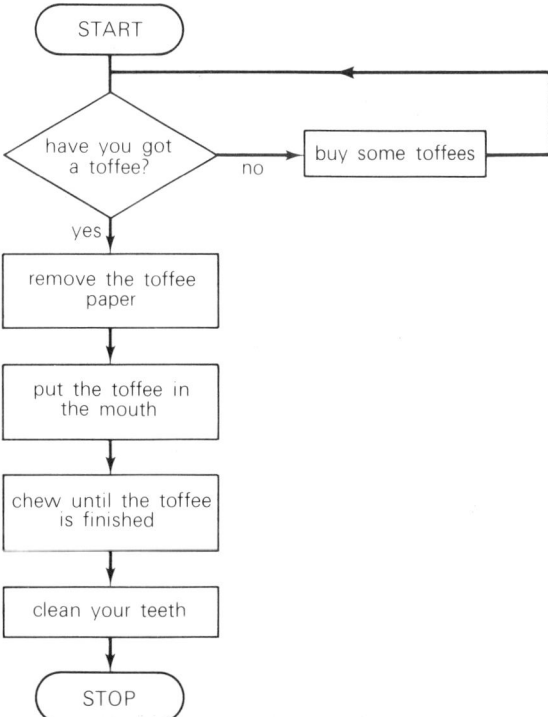

OLD ENGLISH *flowan*, flow.
LATIN *carta*, *charta*, a paper.

FLUID

Capable of flowing. Milk, water and petrol are examples. Gases flow and are therefore fluids. In the **metric system** fluids are generally measured in **litres** or **fractions** of a litre. (See DECILITRE, CENTILITRE and MILLILITRE.)

FOCUS

Special **curves** called **conic sections** can be defined as follows: All **points** on a curve are such that the **ratio**

$$\frac{\text{distance from a fixed point}}{\text{distance from a fixed line}} \text{ is } \textbf{constant}.$$

The fixed point is called the focus. If the ratio is **less than** 1 the curve is an **ellipse**. If it **equals** 1 the curve is a **parabola**. If it is **greater than** 1 the curve is a **hyperbola**.

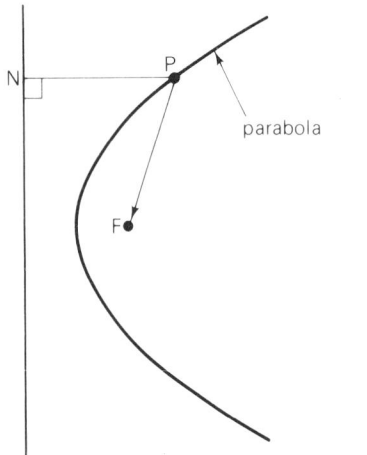

F is the focus of the parabola. For any point P on the curve PN = PF.

FOOT

1 A **unit** of **length** now replaced by units of the **metric system**. It is approximately 30.48 **centimetres**. Originally the length of a foot and hence it varied from person to person until a standard length was agreed. A foot was divided into 12 **inches**.

2 The **point** where a **perpendicular line** meets another line or **plane**. The **diagram** shows B, the foot of a perpendicular from A to CD.

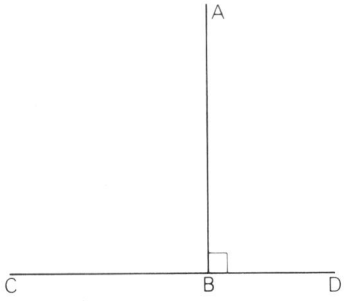

FORCE

Strictly it is that which causes the **acceleration** of a body. Generally used for any push, pull or strain.
Examples: The force of **gravity**. A powerful force is needed to put a rocket into **orbit**.
LATIN *fortis*, strong.

FORMULA

An **equation** containing **symbols** that state the **relation** between two or more **quantities**.
Example: A = ℓ × b (Area = length × breadth)
                C = πd (circumference of a circle = π × diameter).
This is Einstein's famous formula.

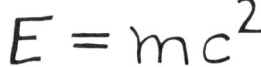

$$E = mc^2$$

LATIN *forma*, form.

FORTNIGHT

Fourteen days. Two **weeks**.

FOUR COLOUR PROBLEM

A **map** drawn on a **plane** is to be coloured so that if two countries have a common **boundary** they are coloured differently.

No one has produced a map requiring more than four colours. Four colours are also sufficient if the countries are drawn on the **surface** of a **sphere**, **cube** or any **solid** that does not contain any holes.

A **torus** (doughnut **shape**) may require seven different colours.

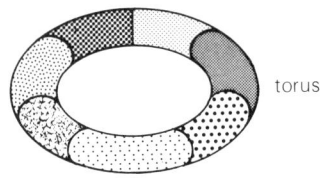

torus

FRACTION

1 The result of **division** (**quotient**). $3 \div 4$ is written $\frac{3}{4}$. $\frac{3}{4}$ is the result of dividing 3 into 4 **equal** parts.
Example: 5 cakes are shared equally by 7 people. Each gets $\frac{5}{7}$ of a cake.

2 A **ratio**. Written as $\frac{3}{4}$ or $3:4$.
Example: John has 6 sweets and Mary has 8 sweets.
Their sweets are in the ratio of 6 to 8.
This can be written as $\frac{6}{8} = \frac{3}{4}$ or $6:8 = 3:4$.
$3:4$ is read as the ratio three to four.

3 As a **number** compared to 1. $\frac{3}{4}$ means 1 is divided into 4 equal parts $\frac{1}{4}$ then 3 of these are taken $(3 \times \frac{1}{4} = \frac{3}{4})$. A fraction can be written as $\frac{a}{b}$ where a and b are numbers.

a is called the **numerator** and b the **denominator**.
(See COMMON FRACTION, DECIMAL FRACTION, EQUIVA-
LENT FRACTIONS, IMPROPER FRACTION, PROPER FRAC-
TION, UNIT FRACTION.)
LATIN *fractum*, to break.

FRANC

A coin in France, Belgium, Switzerland and other
countries.
Originally a 14th century gold coin. It was made of silver
from about 1575 and changed again in 1795. The present
French coin dates from 1960 and is worth 100 centimes.
The early coins were marked 'Francorum Rex' which
means King of the Franks. From this the word franc was
derived.

FREQUENCY

The **number** of **times** an event or **quantity** occurs.

FREQUENCY DISTRIBUTION TABLE

A **table** showing how often each event or **quantity** occurs.
Example:
The marks a class scored in a test are illustrated in this
frequency distribution table.

marks		frequency
100-90	III	3
89-80	ЖН II	7
79-70	ЖН ЖН	10
69-60	IIII	4
59-50	ЖН	5

FREQUENCY POLYGON

A **broken line graph** representing a **frequency distribution**.
It is generally used to describe the **region** contained by the
graph lines, the **horizontal axis** and the **vertical** lines from
the first and last **values** to that axis. In the example below
the heavy lines show the frequency polygon.

FRICTION

The opposing **force** when two bodies are in contact and
one moves or tends to move relative to the other.
LATIN *frictum*, to rub.

There is friction between the tyres and the ground.

FRUSTUM

1 That part of any **solid** contained between two **parallel
planes** that cut through the solid. Particularly applied to
when the solid is a **cone** or **pyramid**.

2 The **term** is also used for the part between any two
planes even if they are not parallel. See examples below.

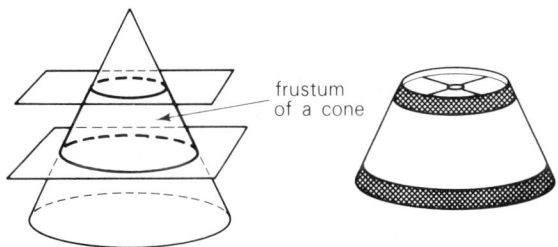

frustum
of a cone

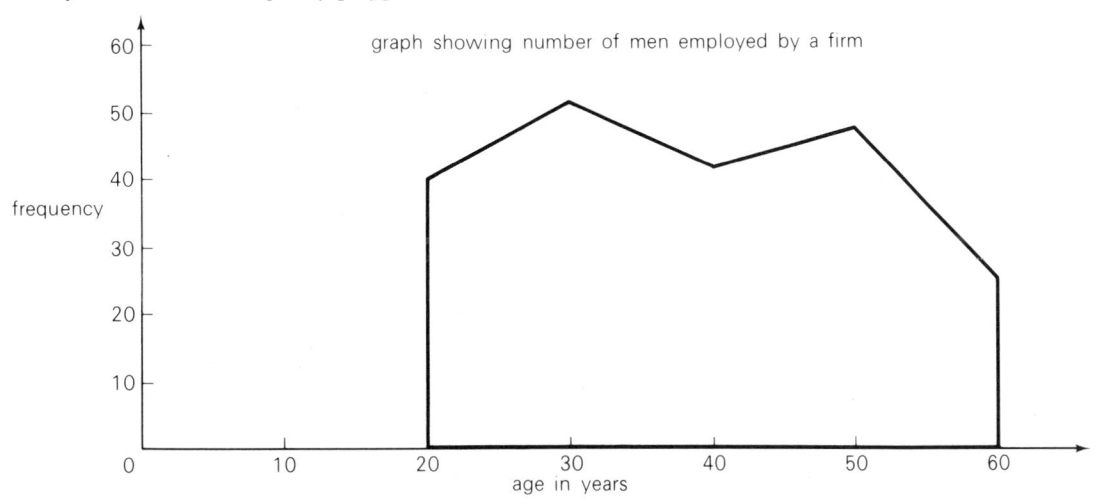

graph showing number of men employed by a firm

frequency

age in years

FULCRUM

The **point** about which a **lever** turns. See picture above.

FUNCTION

A **mapping** of one **set** (the **domain**) to another set (the **range**) so that to each **element** of the domain there **corresponds** only one element of the range. The function is frequently expressed as a set of **ordered pairs**.

Example: (1, 3), (2, 5), (6, 8), (7, 12).

1, 2, 6, 7 form the domain and 3, 5, 8, 12 the range.

1, 2, 6 and 7 each correspond to only one element in the range.

The ordered pairs (1, 3), (2, 5), (2, 6) are *not* a function as 2 corresponds to both 5 and to 6.

A function can be defined in **terms** of a **relation** between two sets such that to any element in the domain there corresponds only one element in the range.

A relation between two sets x and y is a **function** if each element of x is associated with only one element of y.

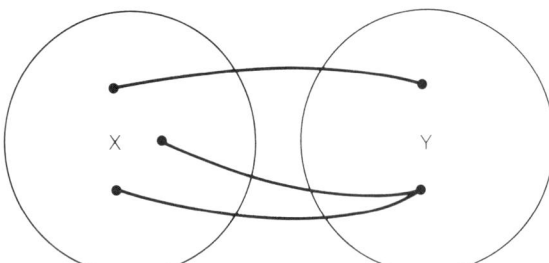

If x is an element of X and y is an element of Y we write $y = f(x)$. This is read as 'y is a function of x'.

Note that one element of Y can be associated with two different elements of X.

FURLONG

A **unit** of **length** in use before the **metric system** was introduced into the United Kingdom.

A furlong was $\frac{1}{8}$ **mile** or 220 **yards**.

OLD ENGLISH *furlang*, length of a furrow (furrow long). A furlang was the length ploughed by an ox before it was given a rest.

G

g

1 The **abbreviation** for the **acceleration** due to gravity.

2 The abbreviation for **gram** (gramme) or grams (grammes).

GALILEO, 1564–1642

A famous Italian scientist and mathematician. He is especially well known for his work on telescopes, how the solar system moves and the movement of bodies acted on by the **force** of **gravity**.

GALLON

A **unit** of **liquid measure** now largely replaced by the **litre**.
A gallon can be expressed in **pints** and **quarts**.
1 gallon = 4 quarts = 8 pints.
A gallon of water is approximately 4.547 litres.

GAUGE

1 To **estimate**.
Example: Gauge the **number** of cupfuls needed to fill a flask.

2 An **instrument** or apparatus for estimating or measuring.
Example: A tyre gauge indicates the **pressure** of the tyre.

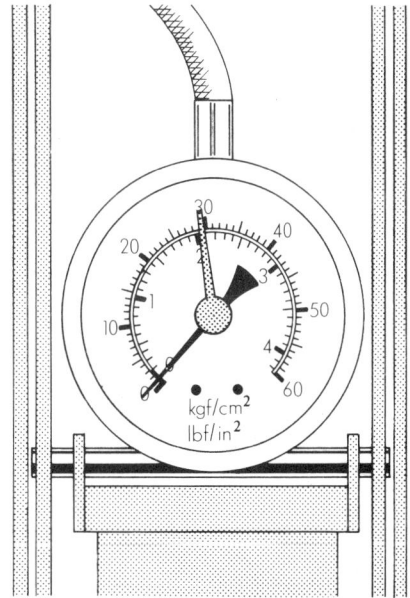

3 The distance between rails on a railway.
OLD FRENCH *gauge*, a liquid measure.

G.C.F.

See GREATEST COMMON FACTOR.

GEAR

A system for transmitting motion, often consisting of wheels and **levers**.

Example: The gears of a bicycle.

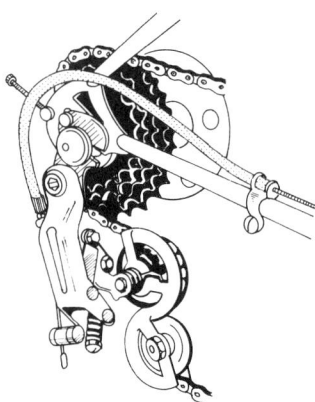

The gear is changed by means of a lever that moves the chain from one cog to another.

OLD FRENCH *gearwe*, preparation.

GENERATOR

1 A **line** or **curve** that moves so as to determine a **surface**.

2 A surface that moves so as to determine a **solid**.

In both cases the movement is frequently a **translation** or a **rotation**.

Examples:

1 An **arc** in the form of a **semicircle** is rotated about a line AB. The arc is the generator. It generates the surface of a **sphere**.

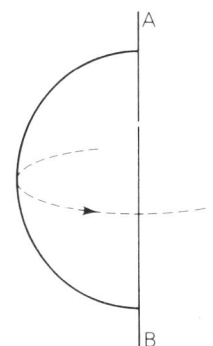

2 A **rectangle**, translated along a line **perpendicular** to its surface generates a **cuboid**. The rectangle is the generator.

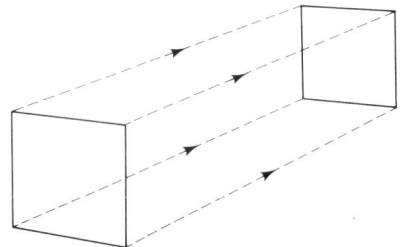

GEO-BOARD

Also called **nail board**, form board or **shape** board. Geo-boards are generally made of wood with nails inserted to form a **grid**.

The nails may be at the **vertices** of **squares** (see illustration) or **equilateral triangles**. In other types the nails are on the **circumference** of a **circle**.

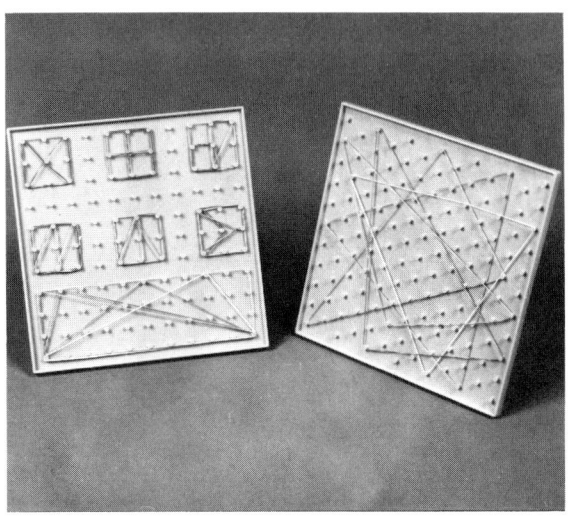

Elastic bands are stretched over the nails to form various shapes whose **properties** can then be investigated. Geo-boards are especially useful for studying **area**.

GEOMETRY

1 The branch of **mathematics** that deals with **solids**, **surfaces**, **lines**, **points**, **angles** and the **relations**, **properties** and **measurements** appropriate to them. Their **positions** in **space** are included in this.

2 In a modern sense geometry includes any mathematical system that is developed from a **set** of **statements** that are called **axioms** or postulates. These are accepted as the starting **point** and not questioned.

Geometry means Earth or land measurement and was used by the Egyptians in **surveying** land and buildings. The Greeks developed this further and **Euclid** systematically arranged all the geometry known at that **time**. This was included in his book called *Elements*. This type of geometry is called Euclidean. The geometry of **conic sections** and other **curves** was added by later Greeks.

Analytic geometry or **coordinate** geometry was developed by **Descartes** in the 17th **century**. Non-Euclidean geometries are advanced forms that do *not* depend upon the assumptions that are made in Euclidean geometry.

GREEK *ge*, Earth; *metron*, **measure**.

GILL

An old **measure** of liquids.

A **quarter** of a **pint**.

GNOMON

1 The part of a sundial that casts the shadow.

2 The remaining part when a **parallelogram** is removed from a **similar** but larger parallelogram.

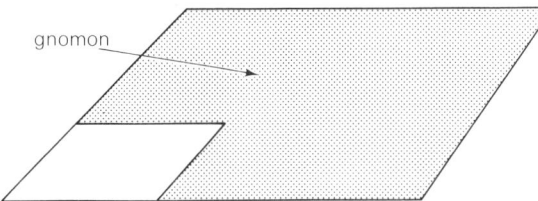

3 Gnomonic **number**. The **odd** number that, added to a **square number**, produces the next square number.

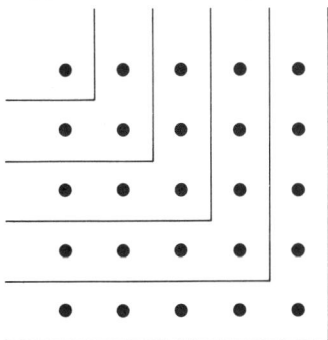

7 added to 9 gives 16.
7 is a gnomonic number.
(The **diagram** shows other gnomonic numbers, 3 and 5.)
GREEK *gnomon*, a carpenter's instrument for making **right angles**.

GOLDEN SECTION

The **division** of a **line** into two parts which are in a particular **proportion** that is frequently found in nature and in art.

One line is approximately 1.618 the length of the other. The **ratio** $\dfrac{1.618}{1}$ or its **equivalent** $\dfrac{1}{0.618}$ is called the golden ratio.

A **rectangle** with sides in the golden ratio is called a golden rectangle. Any golden rectangle can be made into a **square** and a second golden rectangle.

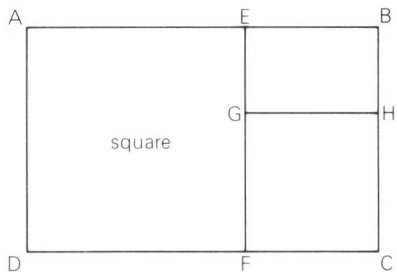

DF is made **equal** to AD and the square AEFD completed. EBCF is a golden rectangle. The process can be repeated GHCF being a square and EBHG another golden rectangle. This can be continued again and again.

GRADIENT

1 The gradient of a **line** is the **measure** of the **amount** of **slope** or steepness. Referred to two **perpendicular** lines such as the **coordinate axis**, it is the **ratio** $\dfrac{a}{b}$.

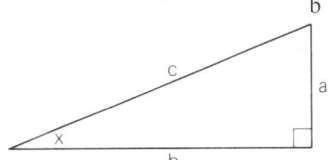

In practical work, such as road **surveying** the ratio $\dfrac{a}{c}$ is used as it is easier to measure c than b.

2 The gradient of a **curve**.
The gradient at P is that of the line touching the curve at P.

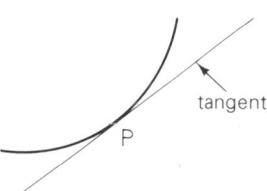

This line is called a **tangent** but is not to be confused with the tangent of an **angle**. In the **diagram** for 1, the tangent of the angle marked x is $\dfrac{a}{b}$.
LATIN *gradus*, a step.

GRADUATE

To mark off for measuring purposes.

Examples: **Rulers** are graduated in centimetres and **millimetres**. We could graduate a **candle clock** in **half-hour intervals**.

GRAM or GRAMME

The **unit** of **mass** in the **metric system**.

Often called the unit of **weight** but should then be called 'gram weight' since weight and mass are *not* the same thing. A cubic centimetre of water has a mass of 1 gram approximately. **g** is the **abbreviation** for gram or grams. gm, gms and g. are not correct.

1000 grams = 1 **kilogram**.

100 centigrams = 1 gram.

1 **ounce** is approximately 28.4 g.

GRAPH

A picture or **diagram** showing the **relation** between **variable quantities**.

1 An algebraic graph or **curve** consists of the **set** of **points** that satisfy a given algebraic **equation** such as $y = 2x^2$ or $\triangle = \square^2 + 2$.

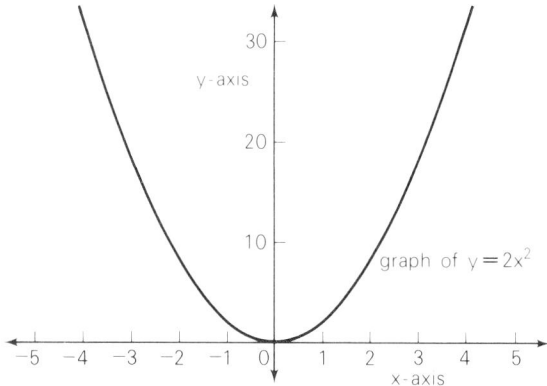

2 See BAR GRAPH; BLOCK GRAPH; BROKEN LINE GRAPH; COLUMN GRAPH; HISTOGRAM; PICTOGRAM; and PIE-GRAPH.

GRAVITY

The **force** with which an object is attracted towards the **centre** of the Earth. This is a special case of a general law stating that any two objects are attracted towards one another. **Newton** established a **formula** for this force F acting on two objects of **masses** m_1 and m_2 that are a distance d apart:

$$F = \frac{k m_1 m_2}{d^2}.$$

k is a **constant**.

LATIN *gravitas*, heavy.

GREAT CIRCLE

The **intersection** of a **plane** through the **centre** of a **sphere** with the **surface** of that sphere.

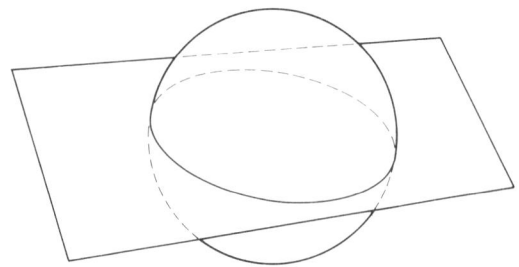

The Earth is not a true sphere as it is flattened near the poles. Because of this the **equator** and the **meridians** of **longitude** are only **approximations** of great circles.

Sailing along a great circle gives the shortest route between two **points**.

OLD ENGLISH *great*, large, important.

LATIN *circulus*, **ring**.

GREATER THAN

More than. Larger than.

The **term** is used when comparing two unequal **numbers**. The **symbol** > denotes 'greater than'.

Example: 5 > 2. Read as 5 is greater than 2.

(See LESS THAN.)

GREATEST COMMON FACTOR

See HIGHEST COMMON FACTOR.

Also called greatest **common divisor**.

GREEK MATHEMATICS

Between 550 BC and 200 BC the Greeks became interested in the reasoning and **logic** of **mathematics**, why results occurred instead of just accepting them. They found contradictions, especially between **arithmetic** and **geometry**. This was largely due to their **number** system, based on their alphabet, for this was too inefficient to enable them to go deeply into the subject. Numbers were thought to have magical properties and mathematicians made money by advising people as to what the future held for them. Some people were actually put to death for giving away these secrets.

The Greeks avoided the contradictions by concentrating on geometry and building this up from simple assumptions (See AXIOM.) The most famous of their books on mathematics were the thirteen written by **Euclid**. The Greeks discovered **properties** of **regular polygons** and **solids**, **irrational numbers**, links with astronomy, ways of measuring the Earth and many other important aspects of mathematics.

Among the most famous Greek mathematicians were **Thales**, **Pythagoras**, **Euclid**, **Archimedes** and **Eratosthenes**.

GREENWICH MEAN TIME (G.M.T.)

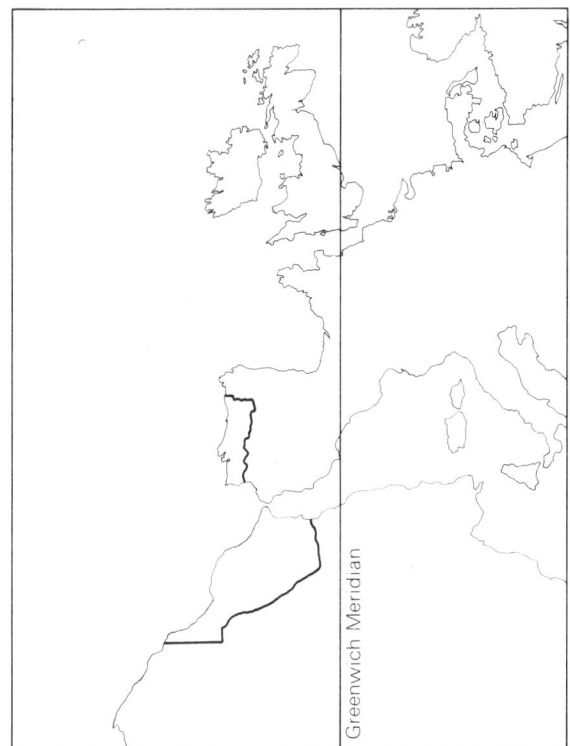

Greenwich Meridian

As the sun moves from east to west, noon as gauged from the **position** of the sun, is different for all places that are not on the same **North**-South line (**meridian** of **longitude**). It would be very inconvenient if towns in a country had different **times** so the world is divided into **time zones**. In any one zone the time is the same at every place.

The British Isles is in the same zone as Portugal, Morocco and part of West Africa. Accurate times were kept at Greenwich Observatory (London) and these took into account the **mean** of the sun's passage over the meridian at Greenwich. Hence the **term** Greenwich Mean Time.

GREGORIAN CALENDAR

The Gregorian **calendar** was introduced by Pope Gregory XIII in 1582. It replaced the **Julian calendar**. A **year** is not exactly $365\frac{1}{4}$ days but is approximately 11 **minutes** 14 **seconds** shorter than this. By 1582 this error had amounted to about 10 days so Pope Gregory decreed that October 4th, 1582 would be followed by October 15th, also that three out of every four **century** years would not in future be **leap years**. Thus 1600, 2000, 2400, 2800, etc. are leap years but 1700, 1800, 1900, 2100, . . . are not.

Riots were caused by this decree as some people thought that they would live ten days less, others objected because debts due to them on the lost days were not paid. The Gregorian Calendar was not accepted in England until 1752.

GRID

Also called a **lattice**.

A grating consisting of two **sets** of **parallel lines**, usually equally spaced along the **direction** of any one set.

Example:

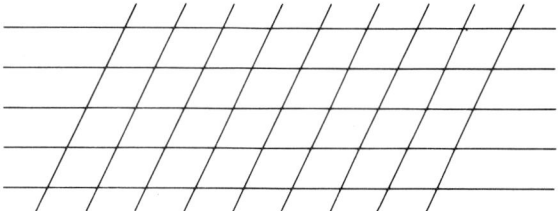

The most useful type of grid is when the sets of lines are at **right angles** to each other and equally spaced. It is then called a **square** grid.

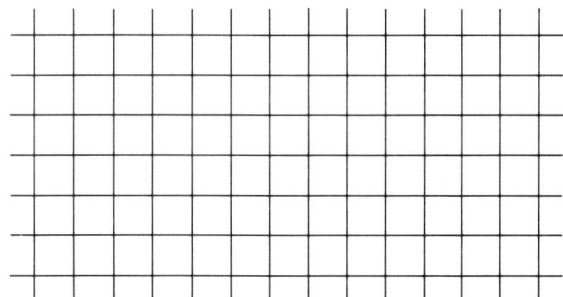

GROSS

144 or 12 **dozen**.

GROSS WEIGHT

The **total weight** of the package.

The weight of the contents alone is called the **net** (or **nett**) weight.

GROUP

1 A mathematical system in which there is a **binary operation** (one combining two **elements**) acting in a **set** and the following apply:

a **Closure.**

b There is an **identity element**.

c Every element has an **inverse**.

d The operation is **associative**.

2 A **collection**. Objects or people that are together.

Example: A group of children were playing football.

GROUPING

Putting together in **sets** with the same **number** in each.
Example: 15 arranged in groups of 5.

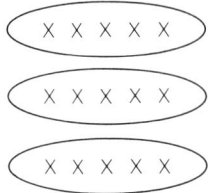

This has links with expressing numbers in **terms** of a **base**.
$15_{ten} = 30_{five}$.
Used in the **quotition** aspect of **division**.
Example: How many 5's in 15? We are grouping in 5's.

GUESS

A guess is an attempt to state a **number** or **quantity** with little or no evidence to base it on.
Example: Guess how many sweets I have eaten this week.
A guess is not the same as an **estimate**. An estimate is based on some available information or earlier experience.

GUNTER, 1581–1626

An English mathematician, Edmund Gunter. Gunter made the first **scale** that is now used for **logarithms**.
Gunter's **chain**: A surveyor's chain with 100 **links**. The chain was 66 feet long and was named after the mathematician mentioned above.

GYRATE

To twist or **rotate**, especially when moving in a path that is like a **spiral**.

H

h

1 **Abbreviation** for **hour** or hours.

2 Abbreviation for **height**. Due to possible confusion with h for hour it is better to use ht for height.

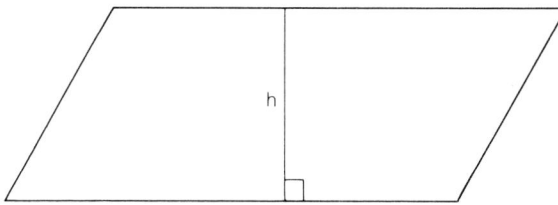

h is the height of the **parallelogram**.

ha

Abbreviation for **hectare**.

HALF

One of two **equal** parts.

Examples:

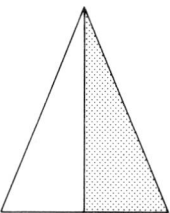

1 Half an **isosceles triangle**.

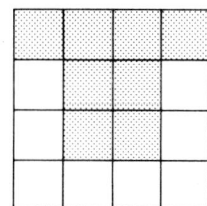

2 Half the **square** is shaded.

(Note: the halves are not the same **shape**.)

3 Half of 19 or $\frac{1}{2}$ (19) = $9\frac{1}{2}$.

HALF-LINE

That part of a **line** on one **side** of a given **point** (P) on the line. It extends to **infinity**.

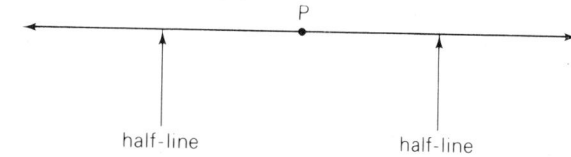

OLD ENGLISH *half*, one of two parts.
LATIN *linum*, flax.
Linen is also made from flax and a stretched linen thread was used to give a **straight line**.

HALF-PENNY

Half of one **penny** in **value**.
Smallest coin in our **currency**.

HALF-PLANE

That part of a **plane** on one **side** of a **line** in that plane.

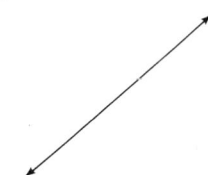

The line and half-plane are thought of as extending to infinity. This is indicated by the arrow-heads.
OLD ENGLISH *half*, one of two parts.
LATIN *planus*, **flat**.

HAND

1 The **width** across a closed hand, now standardised at 4 **inches**. The hand (4 inches) is used in measuring the **height** of horses.

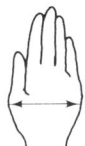

2 A pointer on a **clock** that denotes the **hours** (small hand) or the **minutes** (large hand).

H.C.F.

Abbreviation for **Highest Common Factor.**

HECTARE

1 hectare = 100 **ares**.
1 are is 100 square **metres** so 1 hectare = 10 000 square metres.
Hectare is **abbreviated** to ha.
1 ha is approximately 2.47 **acres**.
GREEK *hekaton*, a hundred.
LATIN **area**, a space or courtyard.

HECTO

Prefix meaning hundred.
1 hectogram (hg) = 100 **grams** (g).
This is approximately 3.52 **ounces**.
1 hectolitre (hl) = 100 **litres** (l).
This is approximately 22 **gallons**.
1 hectometre (hm) = 100 **metres** (m).
This is approximately 109 **yards**.
GREEK *hekaton*, a hundred.

HEIGHT

1 **Altitude.** Distance above the ground or above sea level.
Example: The plane flew at a height of 2000 **metres**.

2 *a* The distance from a **vertex** of a **solid** to the opposite **face**.

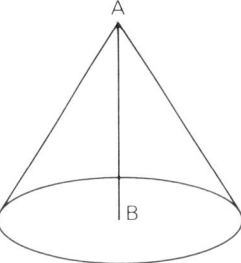

A B is the height of the **cone**.

b The distance from a vertex of a **plane figure** to the opposite **side**.

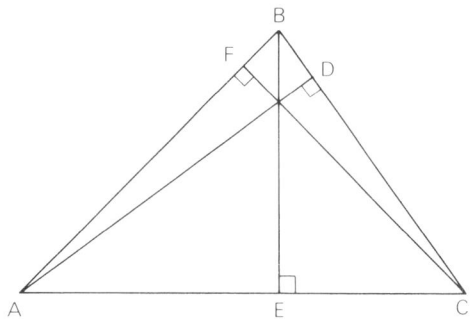

The three **lines** A D, B E and C F are all heights, or altitudes of the **triangle** A B C. It is of interest to note that the three altitudes of a triangle all meet at a **point**.

HELIX

A **curve** that lies on the **surface** of a **cone**, **cylinder** or **sphere**. It cuts a **line** on the **solid** at a **constant angle**.
Examples: A coiled **spring** is generally a helix. Another example is the thread of a screw.

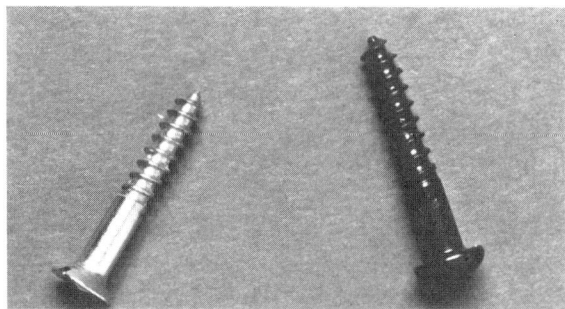

GREEK *helix*, a **spiral**.

HEMISPHERE

A **solid** that is **half** of a **sphere**. The curved **surfaces** are separated by a **great circle**.

hemisphere

HEPTA
Prefix meaning seven.
Heptagon. A **polygon** with seven **sides**.

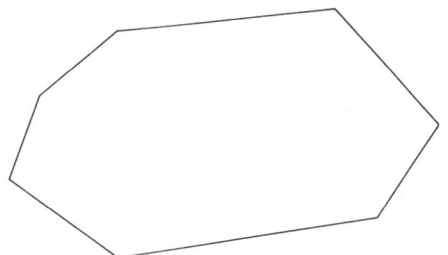

Heptahedron. A **solid** with seven **faces**.
GREEK *hepta*, seven.

HEXA
Prefix meaning six.
Hexagon. A **polygon** with six **sides**.

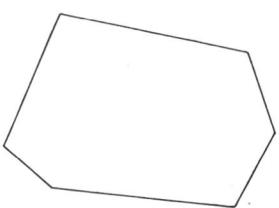

Hexagram. A **figure** made by two **equilateral triangles** intersecting in a symmetrical way as shown.

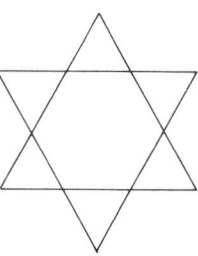

Hexahedron. A **polyhedron** with six **faces**. **Cubes** and **cuboids** are hexahedra.

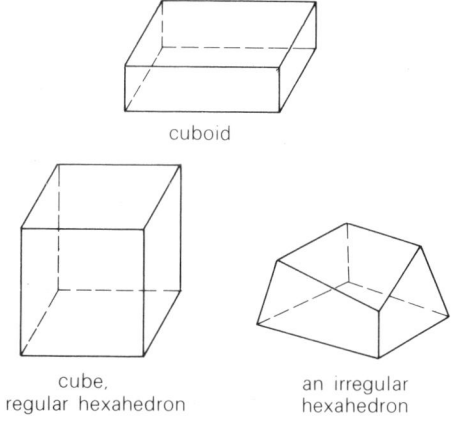

cuboid

cube,
a regular hexahedron

an irregular
hexahedron

Hexomino. A **polyomino** made of six **squares**.

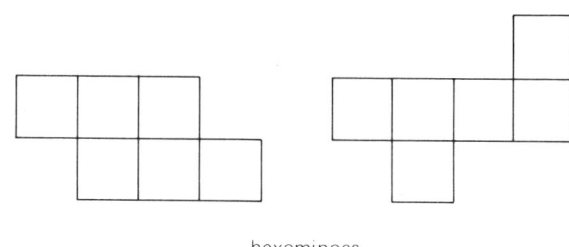

hexominoes

GREEK *hex*, six. *gonia*, **angle**.

HIGHEST COMMON FACTOR (H.C.F.)
Also called **greatest common factor** and **greatest common divisor**.
The highest **number** that divides without a **remainder** into every number in a given **set**.
Example: 4 is the H.C.F. of 16, 36 and 60.

HINDU-ARABIC SYSTEM
Our present **number** system has ten **digits** 0, 1, 2, 3, 4, 5, 6, 7, 8 and 9.
The Hindus used **symbols** for 1, 2, 4, 6, 50 and 200 more than 2000 years ago. By 800 AD they had ten symbols corresponding to our ten digits but it was the combination of **place value** and the use of **zero** as a **place holder** that made **calculations** far easier than they had been.
To appreciate this you could try multiplying two numbers from the **Roman system**.
The Romans did not use place value and zero but we do, so do *not* change the numbers into our system. Multiply XCIV by LIX. These are 94 and 59 in our system so you can easily check your result.
About 700 years ago the Arabs introduced the Hindu system to Europe and this is why it is often called the Hindu-Arabic system.
Leonardo Fibonacci (Leonardo of Pisa) wrote a book in 1202 called Liber Abaci. This explained how to calculate in the Hindu-Arabic system and its advantages over the Roman numbers.

HIRE PURCHASE
A method of buying in which a deposit is paid and the goods delivered. The rest of the money owed is paid in **equal amounts** at agreed **intervals** of **time**.
Example: A television set is bought for £400. £200 is paid as a deposit. The remaining £200 is paid by ten **instalments** of £20 each at monthly intervals.

HISTOGRAM

A **block graph** or **column graph** which has **widths** in **proportion** to the **variable** on the **horizontal axis** and **frequencies** on the **vertical axis**.

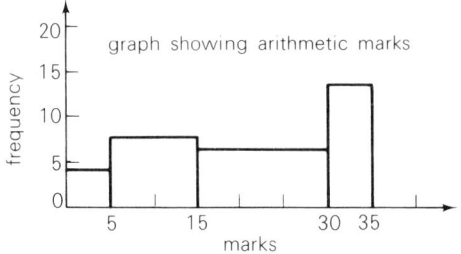

HORIZONTAL

A **line parallel** to the **horizon**.

If the earth is regarded as a perfect **sphere** a horizontal line would be parallel to a line touching the sphere (a **tangent**).

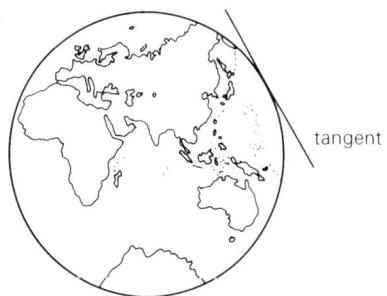

A horizontal line is at **right angles** to a **vertical** line.

HOUR

One twenty-fourth of the **length** of an **average** day. 3600 **seconds**.

HUNDREDWEIGHT

A **mass (weight)** of 112 **pounds**. So called because it used to be 100 pounds. Since the Roman **numeral** for 100 is **C** the **abbreviation** for a hundredweight is cwt.

1 cwt = 112 lb = $\frac{1}{20}$ ton.

1 cwt is approximately 50.80 **kilograms**.

In the United States of America a hundredweight is still 100 lb.

HYPERBOLA

The **curve** formed by a **plane** when it cuts the **surface** of two **cones** positioned as in the **diagram**.

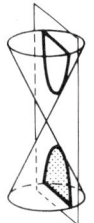

This curve and others formed by cutting a cone are called **conic sections**.

They were studied by the Greeks.

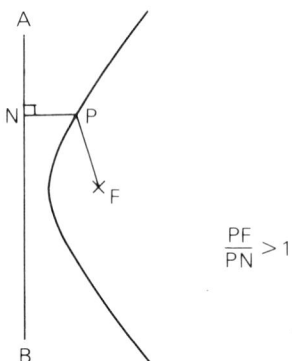

$$\frac{PF}{PN} > 1$$

A hyperbola can also be considered as the **locus** of a **point** (P) such that its distance from a fixed point (F) is **greater than** the distance from a fixed **line** (AB), the **ratio** of the distances being **constant** and greater than 1. The fixed point is called the **focus** and the line the directrix.

HYPOTENUSE

The longest **side** of a **right-angled triangle**. It is always opposite to the **right angle**.

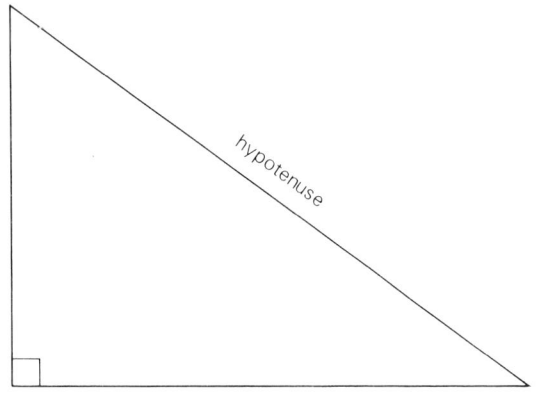

GREEK *hypoteinousa*, stretching under.

I

I
The Roman **numeral** for one.
Fingers were used for early **counting** and **calculating**. I and 1 both resemble a finger.

ICOSAHEDRON
A **polyhedron** with twenty **faces**.
A **regular** icosahedron is one of the five possible **regular solids**. Its twenty faces are **congruent equilateral triangles**.

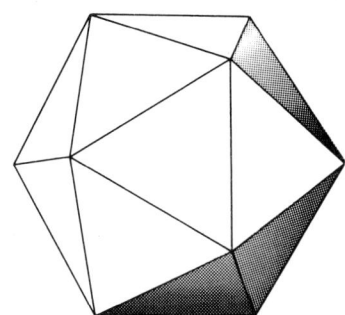

GREEK *eikosa*, twenty; *hedra*, a seat.

IDENTICAL
Exactly the same.
In **geometry** identical **figures** are said to be **congruent**. Identical **sets** have the same **elements** or **members**. Identical sets are more usually called **equal sets**.

IDENTITY
1 An **equation** that is true for all **values**. The **symbol** for identity is ≡.
Example: $(a+b)^2 \equiv a^2 + 2ab + b^2$.

2 Identity **element**. An element of a **set** which, when combined with any other element in that set, leaves it unchanged.
Example: 0 is the identity for **addition**. $3+0 = 0+3 = 3$ (similarly for any other **number**). 1 is the identity for **multiplication**. $a \times 1 = 1 \times a = a$ where a stands for any number.

IMAGE
1 The **reflection** in a mirror.

2 In a **mapping** an **element** in one **set** corresponds to an element in a second set. The element of the second set is called the image.
Example: 6 is the image of 3.

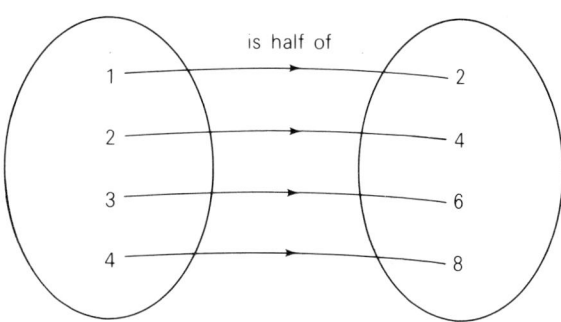

IMAGINARY NUMBER
A **number** whose **square** is **negative**.
You will not have met any numbers that satisfy this as it is advanced work.
$^-3 \times ^-3$ is $^+9$ so $^-3$ is not imaginary.
We need a number such that $\square \times \square = -4$ (or any other negative number). To do this a new type of number is needed which we write as 2i so that $2i \times 2i = -4$.
$4i^2 = -4$, $i^2 = -1$ or $i = \sqrt{-1}$. This number cannot be shown in the way we can show 5 cakes or $\frac{2}{5}$ of a cake. It can only be imagined, yet it is very useful in practical ways such as solving **problems** about electric currents.
(See COMPLEX NUMBER.)

IMPERIAL UNITS
Units of **measurement** used in the United Kingdom and some other countries. Now mainly replaced by the units of the **metric system**.
Imperial units included **inch, foot, yard, furlong, mile, acre, pound (weight** or **mass), hundredweight, stone, ton, pint, quart** and **gallon**.

IMPLICATION

1 A conclusion that follows from what is assumed.
Example: If you do not eat enough food the implication is that you will lose **weight**.

2 A **statement** that depends on certain conditions generally expressed in the form: If ... then ...
Examples:
a If it rains then I will stay at home.
b If $x = 3$ then $x + 1 = 4$.
A statement in the form: 'If p then q' is written as $p \rightarrow q$ and read: p implies q.

IMPROPER FRACTION

A **fraction** in which the **numerator** is **greater than** the **denominator**.
Example: $\frac{8}{5} \begin{smallmatrix} \text{numerator} \\ \text{denominator} \end{smallmatrix}$ $\frac{8}{5}$ is an improper fraction $\frac{3}{5}$ is *not*.
It is a **proper fraction**.
All improper fractions are greater than 1.

INCENTRE

The **centre** of a **circle** drawn inside a **polygon** so as to **touch** each **side**. The circle is called the **incircle**.

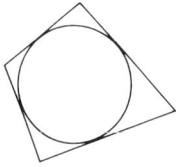

For many polygons it is impossible to draw such a circle, e.g.

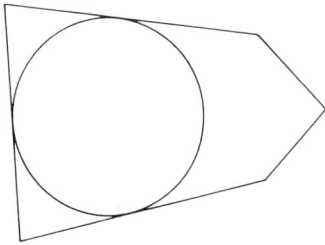

The **term** is particularly used in reference to a **triangle**. A circle can always be drawn to touch the three sides of any triangle, so every triangle has an incentre.

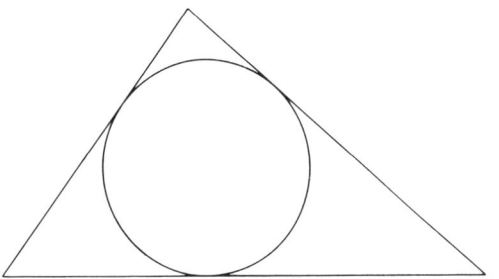

(See INCIRCLE for the method of constructing the incircle and thus finding the incentre.

INCH

A **unit** of **length**. It is approximately 2.54 centimetres. 12 inches = 1 **foot**, 36 inches = 1 **yard**. Units for **area** and **volume** are the square inch (in²) and cubic inch (in³).
OLD ENGLISH *ynce*, an inch. This was derived from the LATIN *uncia*, a twelfth part.

INCIRCLE

A **circle** that touches every **side** of a **polygon**. The **centre** of this circle is called the **incentre**.

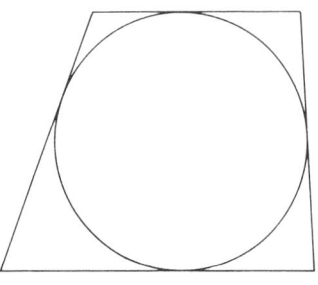

To construct the incircle of a **triangle** we **bisect** any two **angles**. Where these bisectors meet is the incentre. The incircle can then be drawn.

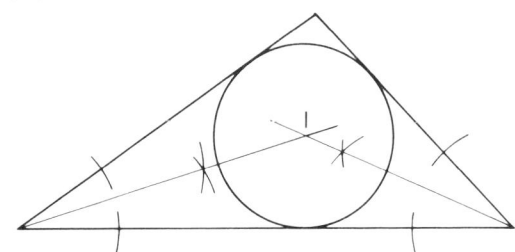

The three bisectors of the angles of a triangle meet at one **point** – the incentre (I).

INCLINE

Slope or **gradient**.
Example: The incline of a hill.
Inclined **plane**. One at an **angle** to a **horizontal** plane.

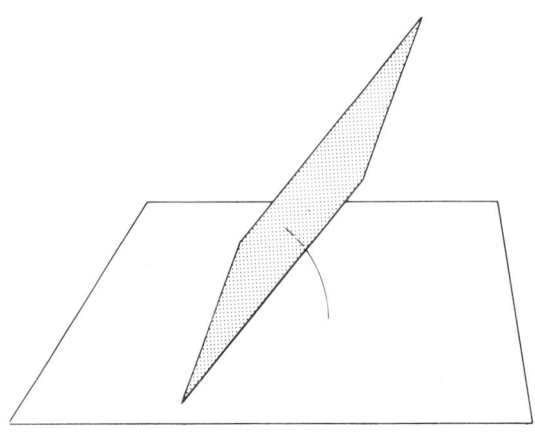

INCLUDE/INCLUSION
Enclose or contain.
Example: The **set** of letters in our alphabet includes the set of vowels. We write $\{a, e, i, o, u\} \subset \{$The letters in our alphabet$\}$. $\subset$ is read as 'is a **subset** of' or 'is included in'. Inclusion is also shown by **Venn diagrams**.

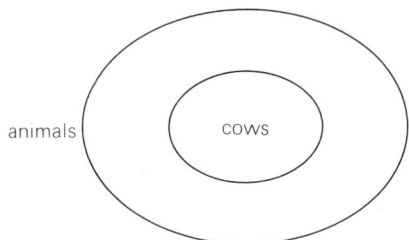

The set of animals includes the set of cows.
This can be written in two ways
a $\{$cows$\} \subset \{$animals$\}$.
The set of cows is included in the set of animals,
b $\{$animals$\} \supset \{$cows$\}$.
The set of animals includes the set of cows.

INDEPENDENT
See DEPENDENT VARIABLE.

INDEX
Plural indices.
The **number** indicating the **power**.
Example: $5^3 = 5 \times 5 \times 5 = 125$. The 3 is the index.
$7^2 = 7 \times 7 = 49$. The index is 2.
When no index is written it is understood to be 1.
Example: $8 = 8^1$.
An index can be any number at all and special meanings are then given.
Examples: $4^{\frac{1}{2}}$ means the **square root** of 4. 5^{-3} means $\frac{1}{5^3}$ or $\frac{1}{125}$.

INDIRECT
Not direct.
For indirect **proportion** and indirect **variation**, see INVERSE PROPORTION.
(See INDIRECT MEASUREMENT.)

INDIRECT MEASUREMENT
Any method applied when it is impossible to **measure** directly.
Examples:
a If you have a **ruler** and a piece of string the **length** of this **curve** can be found.

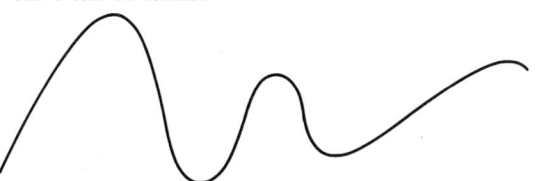

The string is placed along the curve and then the ruler used to find the length of the string. This gives the length of the curve.
b **Temperature** is measured indirectly when we use a **thermometer**. The heat causes the alcohol or mercury in the thermometer to expand or contract and the **amount** of this is read from the **scale**.
c To find the distance of the tower from A. The **angles** at A and B and the distance AB are measured. AC can then be found by a **scale drawing** or by **trigonometry**.

INEQUALITY/INEQUATION
A **statement** that two **quantities** are not **equal**. It can be written with any one of three **symbols**.
Examples: $8 > 7$. 8 is **greater than** 7.
$6 < 10$. 6 is **less than** 10.
$5 \neq 2$. 5 is not equal to 2.
A more complicated example would be $3x + 2y > 7$. The truth of this now depends on the **values** we take for x and y. If $x = 1$ and $y = 1$ the statement becomes $5 > 7$ which is not true. If $x = 5$ and $y = 3$ we have $21 > 7$ which is true. $3x + 2y > 7$ is called an inequation.

INERTIA
The resistance of a body to **forces** that tend to change its state of rest, or of motion.

INFINITE
Unlimited. Without any bounds of **size** or **number**. Not **finite**.
Examples: $0, 1, 2, 3, 4, 5, 6, 7, 8, \ldots$
This **sequence** of **whole numbers** is infinite. (The ... stand for all the missing numbers, $9, 10, 11$ and so on.)
There are an infinite number of **points** on any **line**.
$1 + 3 + 5 + 7 + 9 + 11 + 13 + 15 + \ldots$
The **sum** of all the **odd numbers** is infinitely large. We write the sum as ∞ and say the sum tends to infinity.

INFINITESIMAL

An **amount** that is infinitely small. See **infinite**. If we keep halving a **number** each result gets smaller and smaller. 100, 50, 25, $12\frac{1}{2}$, $6\frac{1}{4}$, ...

If we continue this many, many times the resulting numbers gets nearer and nearer to 0. They become infinitesimal.

INITIAL

At the beginning.

Example:

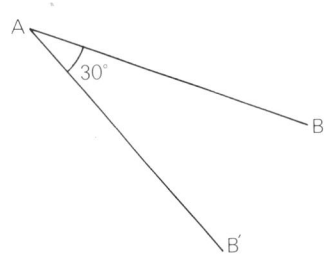

A **line** AB is rotated about A through 30° to the **position** AB′. AB is its initial position and AB′ its **final** position.

INSCRIBE/INSCRIBED CIRCLE

In **mathematics** inscribe means to draw a **figure** inside another so as to **touch** the outer one. A **circle** inscribed in a **triangle** touches its three **sides**.

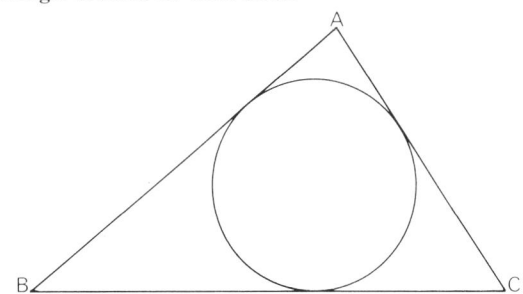

The inscribed circle of triangle ABC.

To find the **centre** of this circle **bisect** the three **angles**. The bisectors will meet at a **point** which is the centre of the required circle. (See INCENTRE and INCIRCLE.)

INSTALMENT

One of a **number** of payments. These are usually **equal amounts** paid at regular **intervals**.

See **hire purchase**.

Example: A man purchases a car and agrees to pay £300 deposit and then twelve instalments of £50. The instalments are to be paid every 2 **months**. He takes 2 **years** to pay and the **total** cost is £900. (£300 + £50 × 12 = £900).

INSTRUMENT

A tool.

Examples: Mathematical instruments: **compasses, protractor, ruler, set square, slide rule, trundle wheel.**

INTEGER

Any of the **positive** or **negative whole numbers** or **zero**, ...
−6, −5, −4, −3, −2, −1, 0, 1, 2, 3, 4, 5, 6, ...
1, 2, 3, 4, ... are called the positive integers.
−1, −2, −3, −4, ... are the negative integers.

INTERCEPT

1 Part of a **curve**, **line**, **surface** or **solid** which is cut off.

Example:

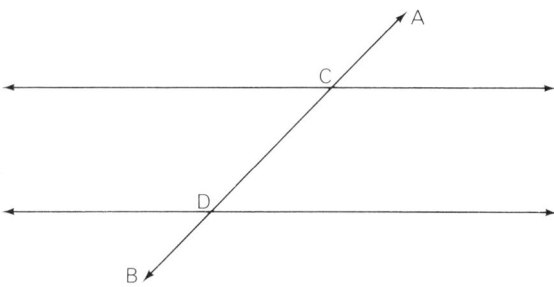

The two **parallel** lines cut AB at C and D. The line CD is the intercept.

2 Also used to mean **intersect**.

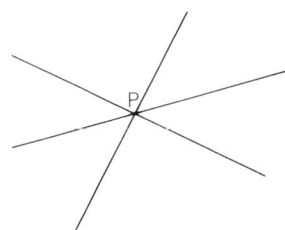

The three lines intercept (or intersect) at P.

INTEREST

A payment for the use of money or goods. The loan is called the **principal**. The interest plus the principal is called the **amount**. **Simple interest** is paid on principal only. **Compound interest** is paid on the principal plus any earlier interest that is due. The **rate** of interest is usually given as a **percentage**. 5 **per cent** per annum means £5 interest is charged on every £100 each **year**.

INTERIOR

Inside. Enclosed.

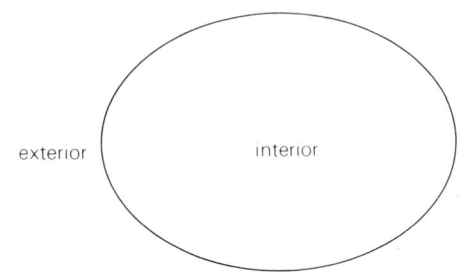

INTERIOR ANGLES
1

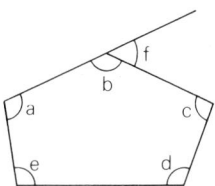

The **angles** on the **inside** of a **polygon**. a, b, c, d and e are interior angles, (f is an **exterior angle**).

2

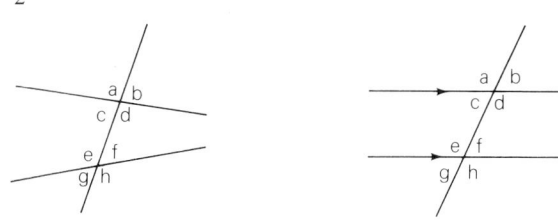

When two **lines** are cut by another (the **transversal**) eight angles are formed. c, d, e and f are called the interior angles. (a, b, g and h are the exterior angles.) If the two lines are **parallel** then c = f and d = e. c and f (also d and e) are called **alternate angles**.

INTERPOLATE
To find a **value** between two others. We may **estimate** this or find it by a **calculation**.
Example: From a **graph** we are told that when
$$x = 2, \quad y = 2.8$$
and when $x = 3, \quad y = 3.2$
We can interpolate to find y when $x = 2.5$. Our best estimate is 3.0 [2.5 is midway between 2 and 3. We therefore find the value midway between 2.8 and 3.2. This is 3.0.)

INTERSECTION of LINES, CURVES, PLANES or SURFACES
Those parts which are **common**.

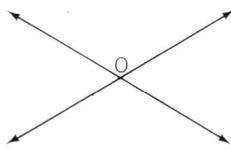

O is the **point** of intersection of the two **lines**.

A and B are the points of intersection of the two **curves**.

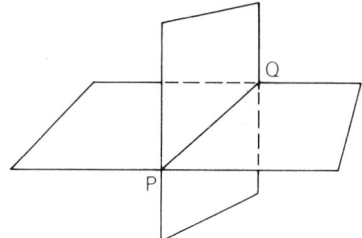

The two **planes** intersect in the line PQ.
LATIN *inter*, between; *sectum*, to cut.

INTERSECTION of SETS
The **set** of **elements** that are common to two or more other sets.

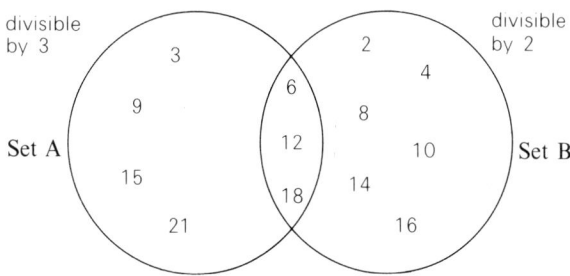

The set {6, 12, 18} is the intersection of set A and set B because 6, 12 and 18 are **divisible** by both 3 and 2. It is written as A ∩ B. This is read as 'A intersect B' or 'A intersection B'.
LATIN *inter*, between; *sectum*, to cut.

INTERVAL
1 The **set** of **numbers** or **points** between two given numbers or points. If the end numbers or points are included the interval is closed. If only one end number or point is included the interval is half-open or half-closed. If neither are included the interval is open.
The ways in which these are indicated are as follows:

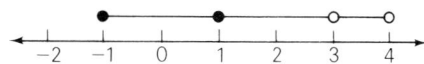

○ denotes 3 is *not* included.
● denotes 1 *is* included.
If x is any number included,

$1 \leqslant x < 3$ Half-open or half-closed interval, 1 is included. 3 is not. Also written [1, 3).
$3 < x < 4$ Open interval. Neither 3 nor 4 are included. (3, 4).
$-1 \leqslant x \leqslant 1$ Closed interval.
Both -1 and 1 are included. [−1, 1].
[and] indicate a value is included. (and) indicate it is not included.

2 An **amount** of **time** between events.

Example: There was a short interval between each shower.

LATIN *inter*, between; *vallum*, a wall. (The interval separates, so does a wall.)

INVERSE

The inverse of an **element** a is an element that, combined with a, gives the **identity**.

For instance in **addition** the identity is 0.

$3 + (-3) = 0$ so the additive inverse of 3 is -3.

For **multiplication** the identity is 1.

$3 \times \frac{1}{3} = 1$ so the multiplicative inverse of 3 is $\frac{1}{3}$.

$\frac{1}{3}$ is the **reciprocal** of 3 and 3 is the reciprocal of $\frac{1}{3}$.

(See INVERSE PROPORTION.)

INVERSE PROPORTION

Also called inverse **variation**, **indirect** proportion or indirect variation.

The **variables** are so related that whatever one is multiplied by the other is divided by that same **number**.

Example: A job required 24 **man-hours** of work. (A man-hour is the work done by one man in one hour.)

number of men	1	2	3	4	6	8	12	24
number of hours they work	24	12	8	6	4	3	2	1

In **algebra** the number of men could be taken as x and the number of hours worked as y. We then have $xy = 24$ (xy stands for x multiplied by y) or $y = \dfrac{24}{x}$. Instead of 24 we might have some other number, say k. Then $xy = k$ or $y = \dfrac{k}{x}$ will apply for any case of inverse proportion.

IRRATIONAL NUMBER

A **real number** that cannot be expressed in the form $\dfrac{a}{b}$ where a and b are **whole numbers**.

(A number that can be expressed as $\dfrac{a}{b}$ is called a **rational number**.)

Example: $\sqrt{3}$ is irrational. It is approximately 1.73 but cannot be expressed exactly as a **decimal** or other **fraction**. π (**pi**) is also irrational. Values such as $3\frac{1}{7}$, 3.14 and 3.1416 are only **approximations**, they are not **exact**.

IRREGULAR POLYGON

A **polygon** that is not **regular**.

To be regular it must have all **sides** the same **length** and all the **angles equal**. Unless both of these conditions are satisfied it is irregular.

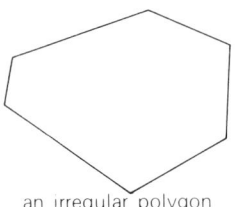

an irregular polygon

ISOBAR

A **line** joining places with the same atmospheric **pressure**.

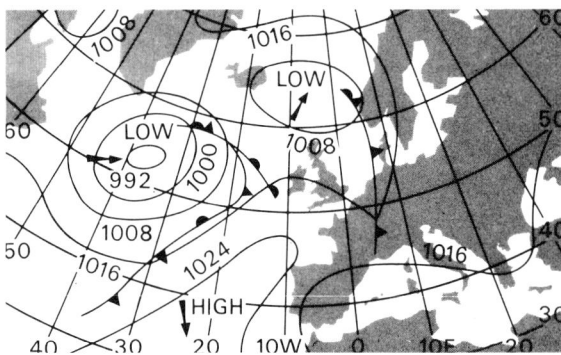

GREEK *iso*, **equal**; *baros*, **weight**.

ISOMETRIC GRAPH PAPER

Paper with a grid of **equilateral triangles** or **dots** which, if joined, would form equilateral triangles.

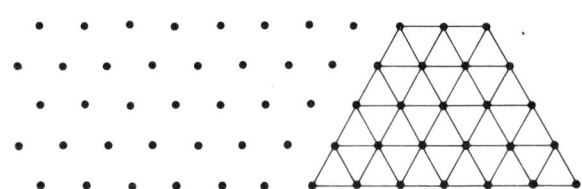

GREEK *iso*, equal; *metron*, **measure**.

ISOMETRY

A **transformation** that leaves the distance between any two **points** unchanged.

Example: . and x are any two points on the **triangle**. They are always the same distances apart.

A translation

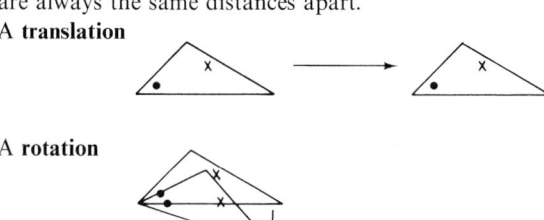

A rotation

A reflection

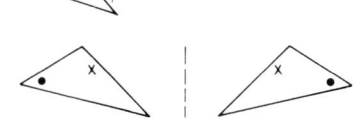

If the triangle was drawn on a rubber sheet and the sheet then stretched the distance between . and x would change. Stretching is *not* therefore an isometry.

GREEK *iso*, **equal**; *metron*, **measure**.

ISOMORPHIC

Two systems are isomorphic if they have the same **structure** or form. This requires

a the **elements** of one system to be put in **one-to-one correspondence** with those of the other system.
and

b **operations** in the one system to have **corresponding operations** in the other.

Example:

+	0	1	2
0	0	1	2
1	1	2	3
2	2	3	4

*	a	b	c
a	a	b	c
b	b	c	d
c	c	d	e

These two systems are isomorphic. + corresponds to * 0 to a, 1 to b, 2 to c, 3 to d and 4 to e.

GREEK *iso*, **equal**; *morphe*, **shape** or form.

ISOSCELES TRAPEZIUM

A **trapezium** in which the **sides** that are not **parallel** are **equal**.

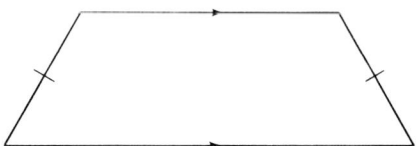

GREEK *isos*, equal; *skelos*, leg.

ISOSCELES TRIANGLE

A **triangle** in which two **sides** are the same **length**.

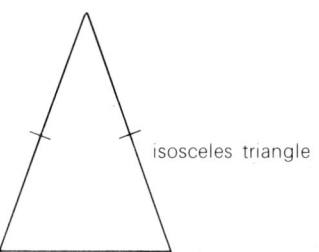

isosceles triangle

Since an **equilateral** triangle has all its sides the same length then it must have two that are the same length. An equilateral triangle is therefore isosceles. Some mathematicians prefer to define an isosceles triangle as 'a triangle with two and only two sides equal'. If this is followed then an equilateral triangle is not isosceles.

GREEK *isos*, equal; *skelos*, leg.
LATIN *tri*, three.

ISOTHERM

A **line** joining places that have the **same temperature**.

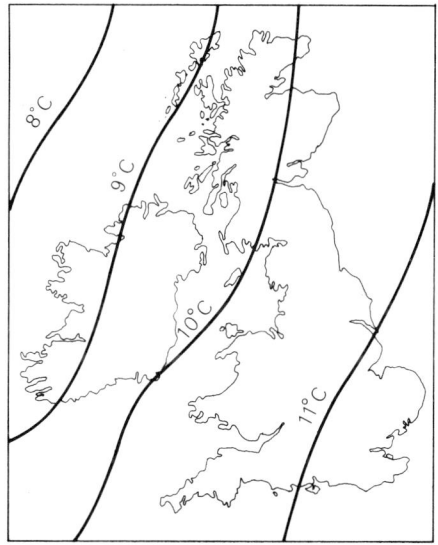

GREEK *iso*, **equal**; *therme*, heat.

J

JOIN
1 A **line** connecting two **points**.

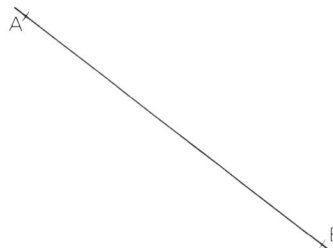

The line AB is the join.

2 An instruction that you are to connect two points.
Example:

×C

Join the points A, B and C so as to form a **triangle**.

JULIAN CALENDAR
In the year 46 BC Julius Caesar abandoned the system of twelve **equal months** (the lunar **year**) and introduced our present solar year of 365 days with one extra day every four years. (See LEAP YEAR for fuller details.)
LATIN. *Julius*, from Julius Caesar; *calendarium*, an account-book.

K

KILOGRAM or KILOGRAMME
The basic **unit** of **mass**.
1 kilogram (kg) = 1000 **grams** (g). Mass and **weight** do not
mean the same. A kilogram is approximately 2.2 **pounds**.
GREEK *chilioi*, a thousand: *gramma*, a small weight.

KILOLITRE
A **measure** of **volume** or **capacity**.
1 kilolitre (kℓ) = 1000 **litres** (ℓ).
1 litre = 1000 cubic centimetres (cm^3) = 1 cubic
decimetre (dm^3).
The **terms** litre and kilolitre are not used for precise work
such as that required in science. A kilolitre is approximately 220 **gallons**.
GREEK *chilioi*, a thousand.

KILOMETRE
A **unit** of **length**.
1 kilometre (km) = 1000 **metres** (m).
A kilometre is approximately $\frac{5}{8}$ of a **mile**.
GREEK *chilioi*, a **thousand**; *metron*, a measure.

KILOWATT
A **unit** of **power**, or **rate** of work.
1 kilowatt = 1000 **watts**.
1 kilowatt is approximately 1.34 horse-power.
GREEK *chilioi*, a **thousand**.
watt, after James Watt (1736–1819).

KITE
A **quadrilateral** with two **pairs** of **equal adjacent sides**. A
kite can be formed by two **isosceles triangles** with a
common base.

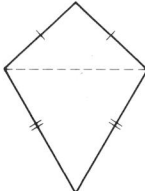

OLD ENGLISH *cyta*, a hawk-like bird.

KLEIN BOTTLE

A **shape** that has only one **surface**, no inside or outside.

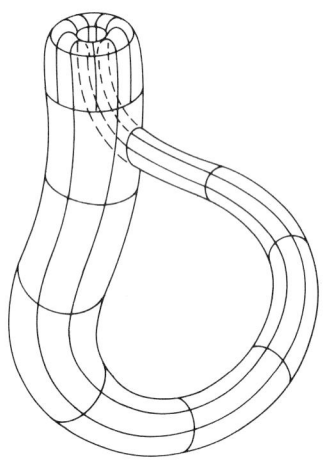

(It is named after the German mathematician Klein (1849–1925).

KNOT

1 A **speed** of 1 **nautical mile** per **hour**.
The name is derived from the fact that a knotted rope was used to calculate the speed of a ship. 1 nautical mile is approximately 6080 feet or 1.15 **statute miles**.

2 A **shape** formed by tying string or some such material and pulling the end tight.

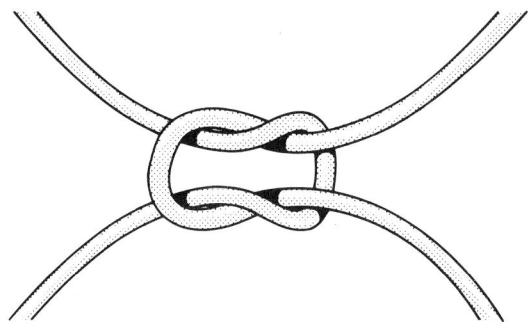

The study of knots is one branch of **topology**.
OLD ENGLISH *cnotta*, knot.

KONIGSBERG BRIDGE PROBLEM

In the town of Konigsberg a favourite walk was along the river bank crossing the river by means of the bridges shown. The bridges joined the two islands as follows:

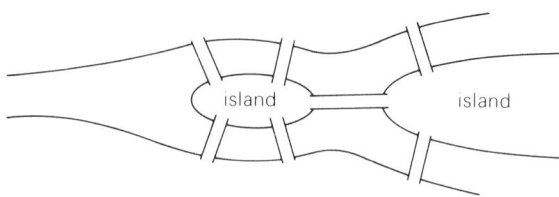

The **problem** was to determine whether or not a person, starting anywhere, could cross each bridge once and once only.
Euler (1707–1783) investigated this problem as part of the branch of **mathematics** called **topology**.

L

ℓ

The **symbol** for **litre**.

Example: One litre is a **thousand** cubic centimetres. The **abbreviation** for this is 1 ℓ = 1000 cm³. Litre is preferably written in full and not abbreviated to ℓ unless it is combined with a prefix as in millilitre, mℓ.

L

The Roman **numeral** for fifty.

The **symbol** L is derived from the Greek letter chi, χ. Chi took various forms throughout the ages ↓, ⊥, ⊥.

L.

L was used for 50 from about the time Christ was born. A less likely suggestion is that it was half of the symbol for 100, C. This was written as ⌈ and the lower half (··⌈··) results in L.

LARGER THAN

See the more correct **term, greater than**.

Example: 8 is greater than 6 can be written as 8 > 6. This is preferred to '8 is larger than 6'.

LATERAL

Side.

Example: **Quadrilateral. A figure** with four sides.

LATIN *lateris*, a side.

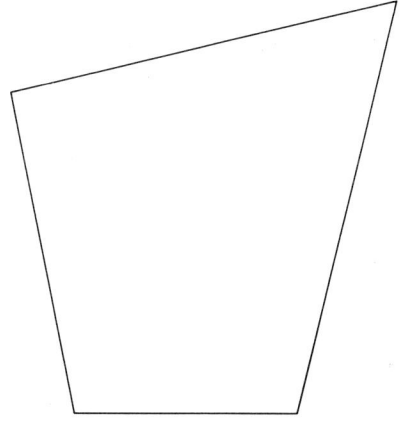

LATIN SQUARE

A **square** with **symbols** in **rows** and **columns**. No symbol appears twice in any one row or column.

0	1	2	3
3	0	1	2
1	2	3	0
2	3	0	1

a	b	c
b	c	a
c	a	b

If the **diagonals** contain each symbol without one being repeated then we have a diagonal Latin square.

LATITUDE

Parallels of latitude are **circles** on the Earth's **surface** with their **centres** on the **axis** joining the **North** and South Poles. The latitude of the **equator** is 0° and that of the North Pole 90°N and of the South Pole 90°S.

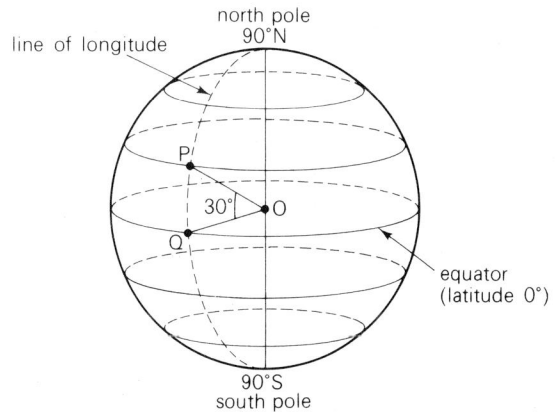

0 is the **centre** of the Earth. In the **diagram** P and Q are on the same **line** of **longitude** (or **meridian**) and Q is on the equator. **Angle POQ measures** the latitude which is therefore 30°.

LATIN *latitudo*, **broad**.

LATTICE

See GRID.

FRENCH *lattis*, a lattice. This came from *latte*, a lath, or thin strip of wood.

L.C.D.

See LOWEST COMMON DENOMINATOR.

L.C.M.

See LOWEST COMMON MULTIPLE.

LEAGUE

A **measure** of distance. It has many different **values**:

A nautical league is $\frac{1}{20}$ of a **degree** (where degree is used as a measure of distance).

3 geographical miles.

Approximately 3.456 **statute miles**.

The Roman league was **equivalent** to approximately 1.376 statute miles.

The French league was about 2.764 miles.

A Gallic league was 1500 Roman **paces**.

FRENCH *ligue*, league. This came from the LATIN *ligare*, to bind.

LEAP YEAR

A **year** in which an extra day is added at the end of February. There are therefore 366 days in a leap year. This adjustment is necessary because the motion of the Earth round the sun takes approximately $365\frac{1}{4}$ days and not 365. Some century years (1800, 1900 for example), are not leap years but apart from these, if the year can be divided exactly by 4 it is a leap year.

Example: $1972 \div 4 = 493$ so 1972 is a leap year. These corrections are slightly inaccurate so every 1000 years *is* a leap year. 2000 will therefore be a leap year.

So called as it is a year in which there is a jump or leap of one day.

OLD ENGLISH *hleapan*, leap; *gear*, year.

LEAST

The smallest or lowest.

Examples: What is the least **amount** you will accept for this bicycle? At least five of the girls had fair hair.

For Least Common Multiple or Denominator see LOWEST COMMON MULTIPLE or DENOMINATOR.

LENGTH

1 The **measure** of distance along a **line** or **curve**.

2 The measure of an **interval** of **time**.

LESS THAN

Not so many as, or smaller than, or fewer than.

The **symbol** for less than is $<$.

Example: $16 < 20$. This is read as 16 is less than 20.

Also see GREATER THAN, $>$.

LEVEL

1 **Horizontal.** The **term** is applied to a **line** or **surface** that is **parallel** to the horizon. A level line or surface is at **right angles** to a **vertical** line, that is one joining a point on the Earth's surface to the **centre** of the Earth.

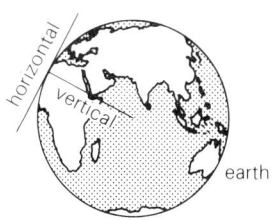

2 All the same **height**.

This is with reference to some given line or surface, generally **mean sea level**.

(See SPIRIT LEVEL.)

LEVER

A rod used to raise heavy objects.

The **point** about which the rod turns is called the **fulcrum**.

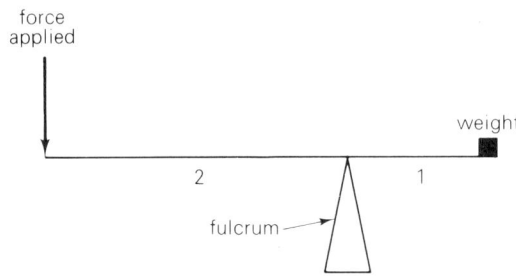

Example: The distances from the fulcrum of the **weight** and **force** applied are in the **ratio** 1 to 2 in this example. The lifting force only needs to be **half** of the weight. If the distances were **equal** (a ratio of 1 to 1) the force and weight would be equal.

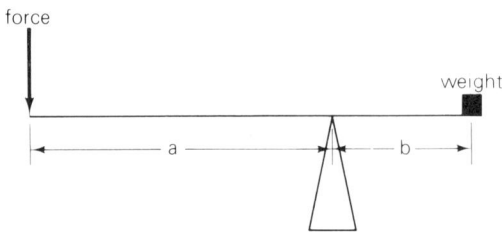

Force $\times$ a = Weight $\times$ b

or $\dfrac{\text{Force}}{\text{Weight}} = \dfrac{b}{a}$

LIBRA

A Roman **pound** abbreviated to lb for the **pound weight** and also giving rise to the pound in money £. The **symbol** £ is derived from the first letter of libra. Originally the pound (money) was one pound-weight of silver.

LATIN *libra*, a **balance**.

LIGHT-YEAR

The distance travelled by light in one **year**. The light-year is used as a **unit** of distance in astronomy. Light travels at approximately 186 000 **miles** (300 000 **kilometres**) per **second** so a light-year is nearly 6 000 000 000 000 miles (9 000 000 000 000 kilometres).

Most stars are more than 100 light-years from Earth. The nearest star, Alpha Centauri is 4.3 light-years away.

LIKE TERMS

Terms in **algebra** that differ in **number** only.
$3x^2$, $19x^2$ and $\frac{1}{4}x^2$ are like terms.
Example: In $5a^3 + 6a^2 + 9a^3 - 4a^2$
$5a^3$ and $9a^3$ are like terms.
$6a^2$ and $-4a^2$ are like terms.

LIMIT

1 The **value** towards which a **series** or **sequence** tends.
Example $1 + \frac{1}{4} + \frac{1}{16} + \frac{1}{64} + \ldots$ The **sum** of the **terms** in this series gets nearer and nearer to $1\frac{1}{3}$.

2 The **position** towards which a **line** moves. A **tangent** is the limiting position of **chords** drawn **parallel** to any given chord.

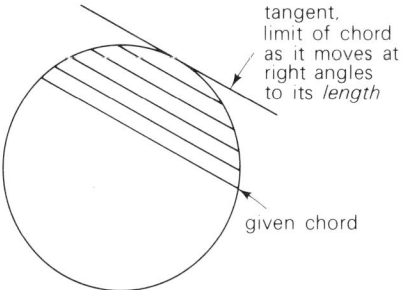

tangent, limit of chord as it moves at right angles to its *length*

given chord

LINE

1 Sometimes used to mean only a **straight line**.

2 More correctly a line may be **straight** or a **curve**. A line, strictly speaking, has **length** but no **width**. If it had width it would be a **rectangle**. But we cannot draw a line without giving it some width no matter how finely we try to draw it. A line is therefore a mathematical idea or **concept**.

Euclid thought of a line as the extension of the shortest path between two points, say A and B.

The correct name for the part of the line between A and B is **line segment**. A line extends infinitely in both **directions** and the arrows are added at the end to show this.
OLD ENGLISH *lin*, flax.
LATIN *linum*, flax. Linen was made from flax and a stretched linen thread was used as a straight line.

LINEAR EQUATION

An **equation** in which the **variable terms** (such as x and y) are of the first **degree** (that is do not have **powers** above 1). *Example:* $y = 4x + 2$ and $3y + x = 5$ are linear equations. If y is plotted against x the **graph** is a **straight line**.

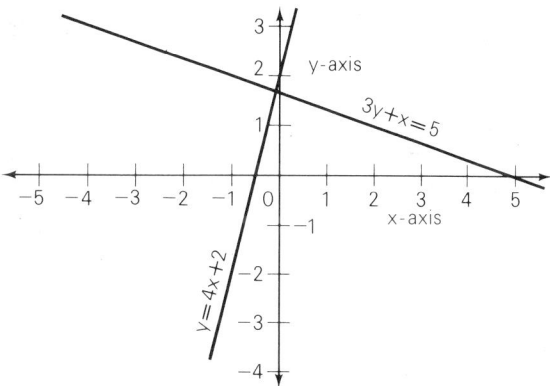

LATIN *linum*, flax (see LINE); *aequalis*, **equal**.

LINEAR MEASUREMENT

The **measurement** of **length**.
This includes the length of **lines** and **curves**. The standard S.I. (**Système internationale d'unités**) **unit** for length is the **metre**.

For longer distances we use the **kilometre**. Short lengths are generally measured in centimetres or millimetres. The most common **imperial units** for linear measurement are **inch**, **foot**, **yard** and **mile**.
LATIN *linum*, flax (see LINE); *mensura*, a measure.

LINE GRAPH

1 A **graph** or chart in which the bars of a **bar chart** are replaced by **lines**.

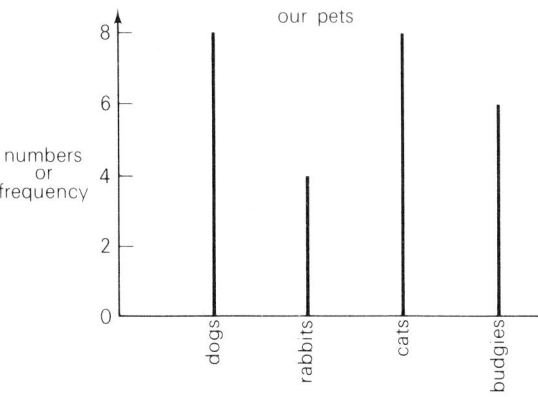

2 Another name for a **straight line graph**.
Example:

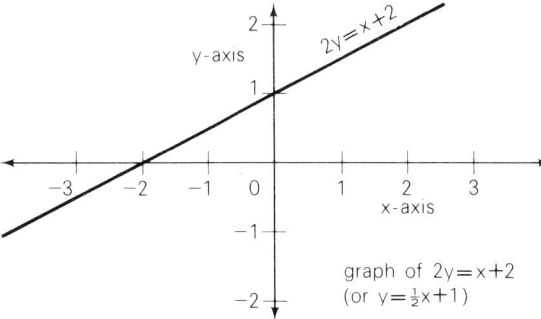

graph of $2y = x + 2$
(or $y = \frac{1}{2}x + 1$)

LATIN *linum*, flax (see LINE).
GREEK *graphe*, something written.

LINE SEGMENT

That part of a **straight line** that contains two **points** and all the points between them

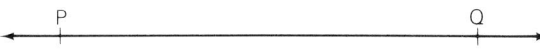

The line segment PQ is that part between P and Q together with the two points P and Q.
LATIN *linum*, flax (see LINE); *segmentum*, cut off.

LINE SYMMETRY

See BILATERAL SYMMETRY. Also called mirror symmetry.
LATIN *linum*, flax (see LINE).

LINK

A **length** that is one hundredth part of the surveyor's or **Gunter's chain**. (1 chain = 22 **yards**). So called because the chain was made of a hundred metal links each of them 7.92 **inches**, (approximately 20.1 **centimetres**).

LIQUID MEASURE

A system for measuring liquids.
In the **metric system** the **litre** is the **unit** for everyday use.
In the **imperial system** the main units are **pints**, **quarts** and **gallons**.
1 litre = 1.7598 pints, that is a little more than $1\frac{3}{4}$ pints.

LITRE

A measure of **capacity** in the **metric system**. It is used for liquids and gases.
It is 1000 cubic **centimetres** (1000 cm^3) or 1 cubic **decimetre** (1 dm^3).
A **litre** is a little more than $1\frac{3}{4}$ pints (1.7598 pints). The **term** is not used in scientific work. (See ℓ.)

LOCUS

The **set** of **points** that satisfy certain conditions. Can also be regarded as the path traced out by a point, **line** or **surface** that moves under stated conditions.
Examples: The locus of a point that is always the same distance from A as it is from B is a **straight line perpendicular** to AB and at **right angles** to AB.

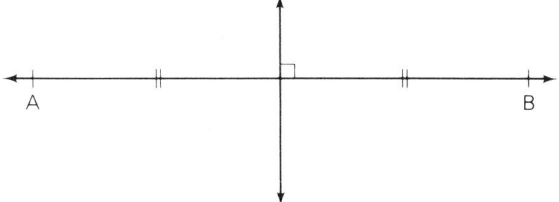

The locus of a line that moves at right angles to itself is a **rectangle**.

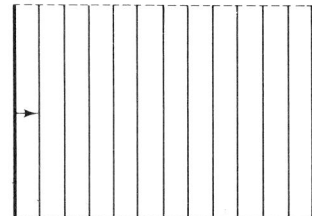

The **right-angled triangle** ABC is **rotated** about AB. The locus of AC is the curved surface of a **cone**.

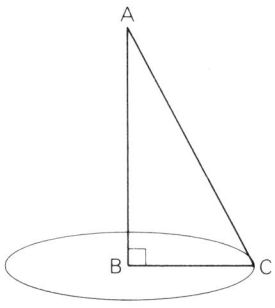

The plural of locus is loci.
LATIN *locus*, place.

LOGARITHM

The **index** or **power** to which a **number** (called the **base**) is raised to give a required number.

Examples: 2 is the base and 16 the required number.
$2^4 = 16$ so 4 is the logarithm of 16 in base 2.
We write $\log_2 16 = 4$.
$\log_{10} 1000 = 3$ (Because $10^3 = 1000$).
For most logarithm **calculations** base 10 is used.
Logarithms enable us to simplify many calculations. **Multiplication** can be replaced by **addition** and other complicated calculations are made much easier. **Tables** are used to find the logarithms of numbers.

Example:

	Number	log
13.83×78.9	13.83	1.1409
$= 1091$ approximately	78.9	1.8971 add
	1091	3.0380

Logarithms were invented by John **Napier**, a Scot in 1594 but his book on the subject did not appear until 1614.

LOGIC

The science of reasoning correctly. Rules can be established, expressed as **symbols** and **calculations** then made in a form that is somewhat like school **algebra**, for instance $[A \underline{v} B]' \longleftrightarrow [A' \underline{v} B]$.

LOGIC BLOCKS

A **set** of wooden or plastic blocks that help to develop logical thinking.

There are many different sets available and one is now described.

There are 3 colours (red, blue, yellow), 4 **shapes** (**triangle, square, circle, rectangle**), 2 thicknesses (thick, thin) and 2 **sizes** (large, small). The blocks contain all possible **combinations** of these **properties** so that there are 48 blocks altogether $(3 \times 4 \times 2 \times 2)$.

For example there is a thick, large, red triangle, a thin, small, blue square and so on. There are many different activities that the blocks can be used for. Among the more important are sorting into sets and combining properties. The popularity of the blocks is largely due to Professor **Dienes**.

LONG DIVISION

A method of dividing

Example:
```
     36 ) 789
          720    20
          ___
           69
           36     1
           ___    __
           33     21
```

$789 \div 36 = 21$ **remainder** 33.

This method can then be shortened to
```
            21
     36 ) 789
          72
          ___
           69
           36
           ___
           33
```

Compare this with **short division**.

LONGITUDE

The **measure** in **degrees** between a **meridian** and the Greenwich Meridian.

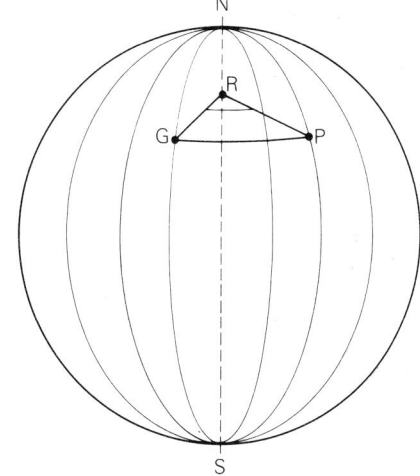

G is Greenwich.

R is in the Earth, on the **line** joining the **North** and South Poles.

Angle GRP measures the longitude of all **points** on the meridian NPS.

Longitude can be up to $180°$ E or $180°$ W of the Greenwich Meridian, $0°$.

Example: P has longitude $70°$ E since the **angle** PRG is $70°$.

LATIN *longitudo*, **length**.

LONG MULTIPLICATION

A form of **multiplication** in which two or more **partial products** are obtained and these are then added to give the answer.

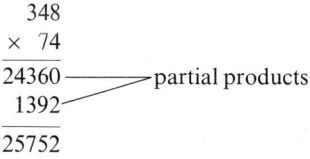

```
       348
     ×  74
     _____
     24360 ———————→ partial products
      1392
     _____
     25752
```

In the example 348 has been multiplied by 70 and by 4. When only one **multiplier** is involved the method is sometimes called short multiplication. This multiplier may, however, have more than one **digit** as the following example shows:

136	Short multiplication. 136 has been
× 12	multiplied directly by 12.
1632	

136	Long multiplication. 136 has been
× 12	multiplied by 10 and by 2.
1360	
272	
1632	

LOOP
A closed curve.

LOSS
The **amount** paid for an article **minus** the amount it is sold for. The loss is therefore **cost price** – **selling price** or C.P. – S.P.

Example: I buy a book for 70p and sell it for 50p. Loss = C.P. – S.P. = 70p – 50p = 20p.

(See PROFIT.)

LOWEST (LEAST) COMMON DENOMINATOR (L.C.D.)
The **least** or lowest **number** that can be divided by the **denominators** of a given **set** of **fractions** without leaving a **remainder**.

Example: $\frac{1}{3}$, $\frac{1}{2}$, $\frac{5}{6}$.

The L.C.D. is 6 since 6 is the lowest number that can be divided by 3, 2 and 6 without leaving a remainder. When fractions are added or subtracted they can first be expressed with the same common denominator.

$\frac{3}{5} + \frac{7}{10} + \frac{1}{3}$.

L.C.D. of 5, 10 and 3 is 30.

$\frac{18}{30} + \frac{21}{30} + \frac{10}{30} = \frac{49}{30} = 1\frac{19}{30}$.

LOWEST (LEAST) COMMON MULTIPLE (L.C.M.)
The **least** or lowest **number** which is a **multiple** of a **set** of given numbers.

Example: Given 6 and 8 then multiples of 6 are 6, 12, 18, 24, 30, 36, 42, 48, 54 ... Multiples of 8 are 8, 16, 24, 32, 40, 48, 56 ... **Common multiples** of 6 and 8 are 24, 48, 72 ...
The least of these is 24 so the L.C.M. of 6 and 8 is 24.

LOWEST TERMS
When the **numerator** and **denominator** of a **fraction** have no **common factor** (other than 1) the fraction is in its lowest terms.

Example: $\frac{11}{18}$ is in its lowest terms but $\frac{16}{18}$ is not. $\frac{16}{18}$ can be **simplified** to $\frac{8}{9}$. The fraction is then in its lowest terms.

LUNAR MONTH
The **time** between a phase of the moon and its next occurrence.

This varies from **month** to month but is approximately $29\frac{1}{2}$ days.

LATIN *luna*, the moon.

OLD ENGLISH *monath*, month; from *mona*, moon.

M

M

The Roman **numeral** for 1000.
There are two possible reasons for its choice:
a It was the initial letter of mille which meant 1000.
b It may have come from Milia (**mile**) which was 1000
    Roman **paces**.
Earlier forms of M are Φ, ◌ and ⊂⊃. A **line** over the top
of a **number** made it 1000 **times** as large so M̄ stood for
1 000 000. We are used to the fact that in the **Roman
system** a small number in front of a larger one denoted
subtraction, IX was 9 IV was 4.
This did not always apply when one of the numbers was
large as the two were then **multiplied**. For instance XM
was 10 000 (not 990 from 1000 − 10). This came from
writing ten milia as XM (ten miles) and so since a milia
was 1000 paces, XM came to represent 10 000.
In the Greek (Alexandrian) number system M represented
40.
LATIN *mille*, a **thousand**.

m

The **abbreviation** for **metre** or **metres**.

M.A.B.

The **abbreviation** for **multi-base arithmetic blocks**.

MACHINE

Any device for doing work.
It may use ropes, **gears**, etc. and the **force** could be
provided by a man's muscles or by a fuel such as petrol,
coal or electricity.
Examples: Cars, **computers**, lawn mowers, **levers**, **pulleys**.

MAGIC SQUARES

An arrangement of **numbers** in the form of a **square** such
that the **total** for each **row**, **column** and **diagonal** is the
same.

Examples:

8	1	6
3	5	7
4	9	2

A magic square of the third **order**.
All the totals are 15.

13	8	12	1
2	11	7	14
3	10	6	15
16	5	9	4

A magic square of the fourth order.
All the totals are 34.

MAGNIFICATION

The **factor** by which a **quantity** has been increased.
Example: An insect of length 1 mm is placed under a
microscope. The magnification makes the insect appear to
be 4 mm long. The magnification factor is 4.
(See DILATATION, ENLARGEMENT.)

MAGNITUDE

The **size**. Used when referring to any **measurement**.

MAJOR

The greater in **number**, **size** or **value**.

Examples:

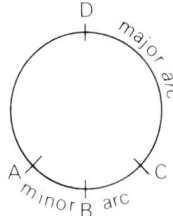

Arc ADC is the major arc.
Arc ABC is the minor arc.

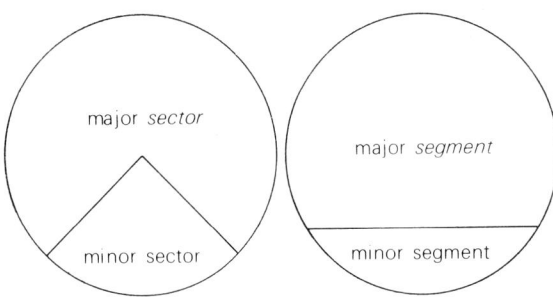

Compare with **minor** which applies to the smaller part.

MAN-HOUR

The **amount** of work 1 man can do in 1 **hour**.

Examples: 3 men work for 8 hours a day. They do 24 man-hours. A job requires 56 man-hours. How many men are needed if it is to be completed in 14 hours? (Answer: 4.)

OLD ENGLISH *mann*, man.

GREEK *hora*, an **interval** of **time**.

MANY-ONE MAPPING

A **mapping** in which more than one **element** (or **member**) of the first **set**, **maps** to the same element in the second set.

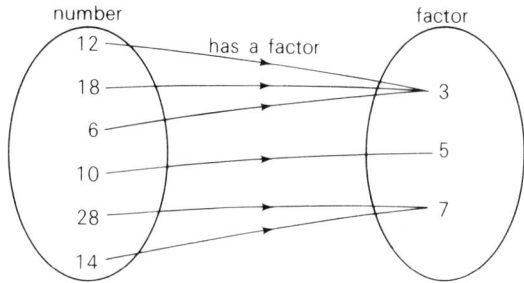

A many-one mapping (or **correspondence**)

MAP. MAPPING

The **operation** of making a **correspondence** between two **sets** such that every **element** (or **member**) of the first set has one, and only one, corresponding element in the second set.

The element in the second set that corresponds to the one in the first set is called its **image**. In the mapping below 3 maps on to 10 so 10 is the image of 3.

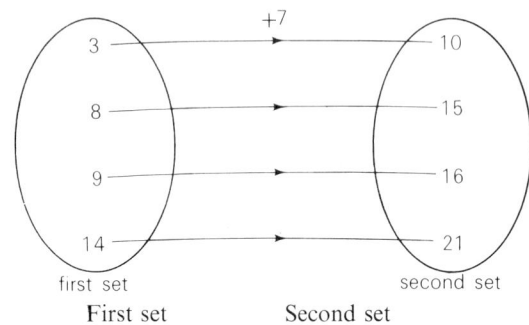

First set   Second set

MAP MEASURER

An **instrument** that **records** the distance on a **map**. A wheel is pushed along a route on the map and the distance can then be read directly from a **dial** on the measurer.

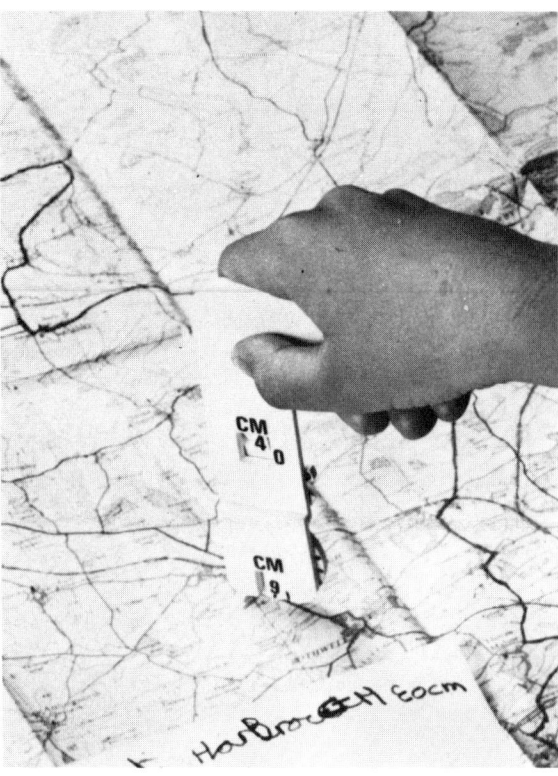

An allowance can be made for the **scale** of the maps by making an appropriate setting on the dial.

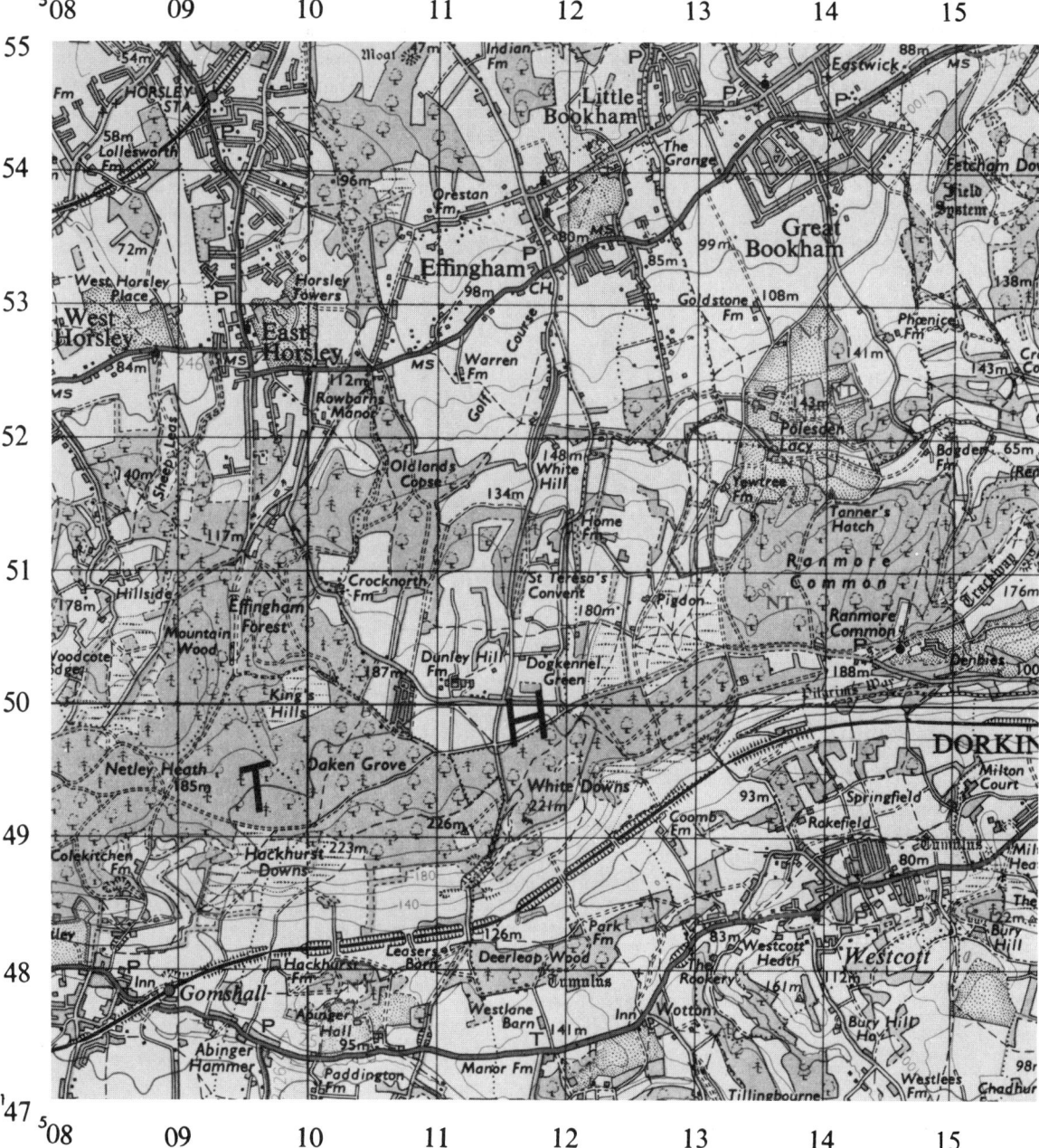

MAP REFERENCE

Letters or **numbers** marked on **grid lines** that cover a map so that places can be easily located and their **positions** given.

Goldstone Farm has a six-figure map reference 135530. The first three figures show the distance east, called the Easting, which is 13.5 but the point is omitted. The last three figures show the distance north, called the Northing, which is 53.0 but the point is omitted. The six figures are written without any spaces – 135530.

MASS

Very simply the **amount** of matter in a body. It is closely linked to **weight** but is not the same. The weight of a body is the **force** with which it is attracted towards the Earth's **centre**. When the body is in **space** it reaches a **position** where the forces due to the attraction of other stars and planets **balance** the force due to the Earth. The body is then weightless but its mass is the same as when it was on the Earth's **surface**. The astronaut has the same mass when on the moon as when on Earth. His weight is far less on the moon because the moon is far smaller than the Earth and its force of attraction is therefore less.

LATIN *massa*, a lump.

MATCH

1 To place in **one-to-one correspondence**.
Example: In two football teams each player can be matched with the corresponding player from the other team, centre-forward with centre-forward, left-back with left-back and so on.

2 To **pair**. This differs from 1 above as cases such as the following show: Children who danced together: (John, Mary), (Jack, Ann), (John, Betty), (Alan, Mary), (Jack, Betty). The children are in pairs but *not* in one-to-one correspondence.

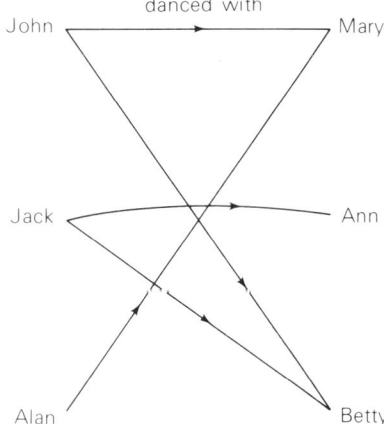

MATHEMATICS

A complicated but highly organised system that has many branches.

At an elementary level there is **arithmetic**, **algebra** and **geometry** but each of these has been extended at higher levels and many new branches added. **Trigonometry, topology, mechanics, dynamics, statistics, probability, analysis** and **logic** are but a few.

There are many possible **definitions** of mathematics but none of them adequately describe such a complex subject. For example: Mathematics is the study of **patterns** and **relations** and the means of representing and communicating them.

MAXIMUM

1 The greatest **value**.
Example: The maximum **temperature** was 20°C.

2 Mathematicians give a special meaning to maximum. It is the greatest **value** in the immediate vicinity (near) but there must be **points** on either **side**, as at A. The value at C is 10 but that at A is only 7, yet A is a maximum point and C is not (as there are no points on one side of C).

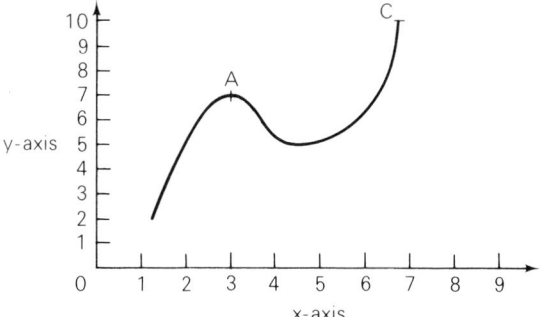

MAXIMUM-MINIMUM THERMOMETER

A **thermometer** that **records** the highest and lowest **temperatures**.

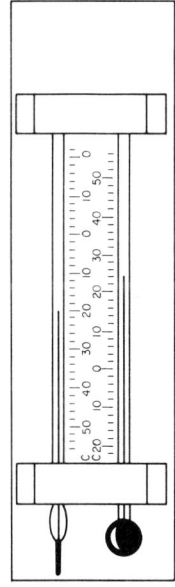

MEAN

Another **term** for **average**. There are three main types:
1 **Arithmetic mean.** The **sum** a set of **quantities** divided by the **number** of them.
Example: 3, 5, 9, 11.

$$\text{Arithmetic mean} = \frac{3+5+9+11}{4} = \frac{28}{4} = 7$$

2 **Median.** The middle **value** when numbers or quantities are arranged in **order**.
Example: 6, 1, 8, 23, 17. In **ascending order** 1, 6, 8, 17, 23. 8 is the median.

3 **Mode.** The most frequent value.
Example: 1, 1, 3, 3, 3, 4, 7, 7, 7, 7, 8, 8. 7 is the mode. Geometric and harmonic means are two others that are used in advanced mathematics.

MEAN SEA-LEVEL

The **height** of the sea changes from **time** to time. The **average** of these heights is called the mean sea-level and this will also change from one place to another.
Ordnance survey maps give heights above the mean sea-level at Newlyn, Cornwall.

MEASURE

1 Also called measurement. The **size** in **terms** of some agreed **unit**. This may be an **arbitrary unit** such as the **number** of pencil **lengths**, cupfuls or **spans**, or it may be a **standard unit** that is recognised and accepted throughout a country or countries. Examples of standard units are **metre**, **second**, **degree** and **gram**.
Example: The measure of the path was 15 metres.

2 The **operation** of finding a **quantity**.
Example: Measure the height of your friend.

MEASUREMENT

Another **term** for **measure**.

MECHANICS

The branch of **mathematics** which studies the effect of **forces** on bodies.
There are two main branches **statics** and **dynamics**. Statics deals with forces on bodies that are at rest and **dynamics** with forces that are in motion.

MEDIAN

1 The middle **value**. One form of **mean**.
Example: 3, 17, 2, 14, 9. When placed in **order** 9 is the middle value or median.

2 A **line** drawn from a **vertex** of a **triangle** to the midpoint of the opposite **side**.

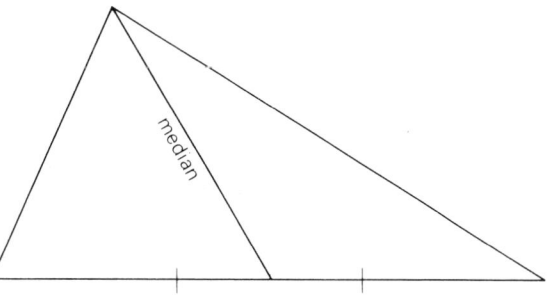

MEDIUM
The middle **value**. For example a person might be of medium **height**.
In **statistics** this is called the **median**.

MEET
The **intersection**.
Examples:

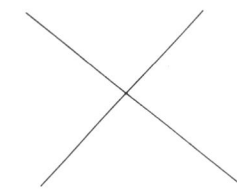

The **point** where **lines** meet or intersect.

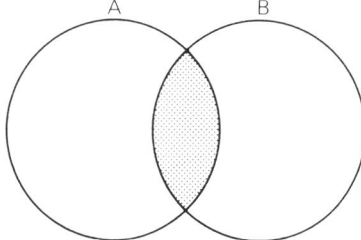

The shaded **region common** to **sets** A and B. It is written A ∩ B and read as 'A intersection B'.

MEMBER
Also called **element**.
A **symbol**, **value** or object that belongs to a **set**.
Examples: a, h, g and k are the members of the set of letters we call the alphabet. 1, 3, 5, 7, 9, 11, 13, . . . are members of the set of **odd numbers**.

MENSURATION
The process of measuring, especially applied to **length**, **area** and **volume**.

MERIDIAN
An **arc** joining the North and South Poles. Meridians are **semicircles** and halves of **great circles**.
Every **point** on the Earth's **surface** lies on a meridian. The **angle** between that meridian and the one passing through Greenwich gives the **longitude** of that point.

METER
An **instrument** for measuring.
Examples: The speedometer of a car. Parking meter.
Meters for measuring the **amount** of gas or electricity used.

LATIN *meta*, a **boundary**.

METRE
The basic **unit** of **length** in the **metric system** and the International System (**Système Internationale d'Unités**). Originally it was one ten-millionth of the **meridian** through Paris, between the North Pole and the Equator. Later it was defined as the distance between two marks on a bar of alloy at the **Weights** and **Measures** Bureau, Severes, France. It is now defined in **terms** of the wavelength of light from a gas, krypton-86. A metre is approximately 39.37 **inches**.
GREEK *metron*, a measure.

METRICATION
The change from using **imperial units** (inches, tons, ounces, etc.) to those of the **metric system** or to S.I. units.
S.I. stands for **Système Internationale d'Unités**.

METRIC SYSTEM
A **decimal** system of **measurement** that is based on ten. It was adopted by France at the time of the French Revolution.
A version of the metric system better suited to science has now been internationally agreed upon. It is the **Système Internationale d'Unités** (S.I.) and is based upon the **metre**, **kilogram** and **second**.

METRIC TON
Also known as a **tonne**.
1000 **kilograms**.
About 0.98 of an imperial **ton**. (An imperial ton is 2240 **pounds**.)

MICRO
Prefix meaning one millionth ($\frac{1}{1\,000\,000}$).
Example: A microsecond is one millionth of a **second**.

MICROMETER

An **instrument** for measuring very small **lengths**, thicknesses or **angles**. A micrometer has two **scales**, one on the barrel (A) and the other on the thimble (B). The **combination** of these enables very **accurate** readings to be made.

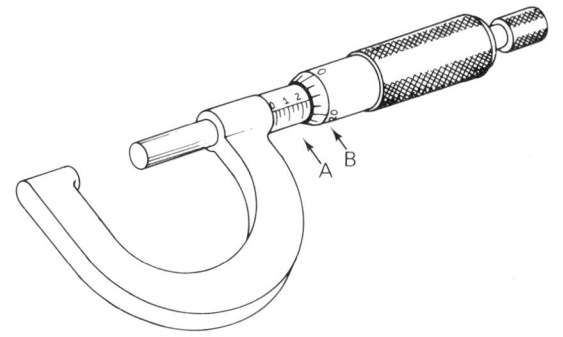

MID

Short for middle
Examples: Midday (Noon). Midpoint.

C is the midpoint of **line** AB.
Midsummer day. June 24th.

MILE

A **measure** of **length**.
See STATUTE MILE and NAUTICAL MILE.
LATIN *mille*, a **thousand**. (The Roman mile was 1000 **paces**.)

MILES PER HOUR

A **measure** of **speed** in the **imperial** system.
The **average** speed in miles per hour is calculated by

$$\frac{\textbf{total distance in miles}}{\textbf{time taken in hours}}$$

Example: A journey of 48 miles takes $1\frac{1}{2}$ hours.

$$\text{Average speed} = \frac{48 \text{ miles}}{1\frac{1}{2} \text{ hours}} = 32 \text{ miles per hour}.$$

MILLI

Prefix meaning one thousandth.

1 milligram $= \frac{1}{1000}$ **gram**.
1 millilitre $= \frac{1}{1000}$ **litre**.
1 millimetre $= \frac{1}{1000}$ **metre**.

LATIN *mille*, a **thousand**.

MILLION

A **thousand** thousand. 1 000 000.

MINIMUM

1 The smallest or **least value**.
Example: The man refused to accept less than £1000 for his car. £1000 was the minimum price.

2 Mathematicians give a special meaning to minimum. It is the smallest value in the immediate vicinity (near) but there must be other points either side, as at B. The value at A is 1 but at B it is 3, yet B is a minimum point and A is not (because there are no points on one side of A).

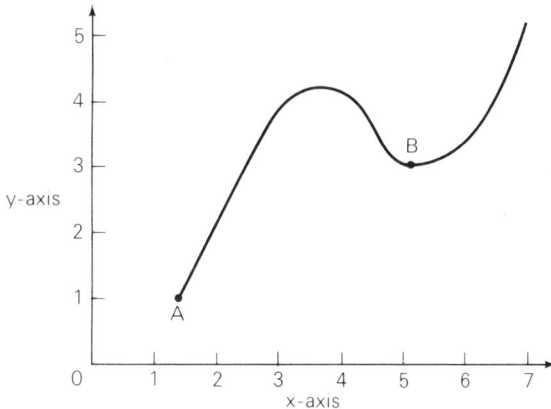

MINUEND

A **number** from which another is subtracted.
Example:

$$\begin{array}{rl} 28 & \text{minuend} \\ -19 & \textbf{subtrahend} \\ \hline 9 & \end{array}$$

MINOR

Smaller.
Especially applied to **arc**, **sector** and **segment**.

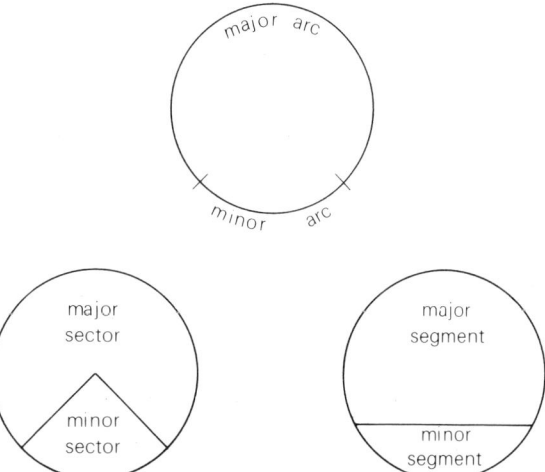

Compare with **major** which applies to the greater or larger part.

MINUS

The name for the **symbol** $-$. It stands for the **operation** of **subtraction**. $7-5$ is read as 7 minus 5 and means 5 is to be subtracted from 7.

When written in a higher **position**, as in $^-3$ it is read as negative three. 3 is a **negative number** and the operation of subtraction is not implied. There is however a close link with -3 which can be thought of as $0-3$ or subtract 3 from 0. Minus was written as m and then carelessly as $\frown$. From this it became $-$.

MINUTE

1 60 **seconds**. $\frac{1}{60}$th of an **hour**.

2 In angular **measurement** $\frac{1}{60}$th of a **degree.**

3 Very small. (Spelt the same as 1 and 2 but pronounced differently.)
Example: The disease was caused by a minute germ.

MIRROR IMAGE

The **reflection** as seen in a mirror.

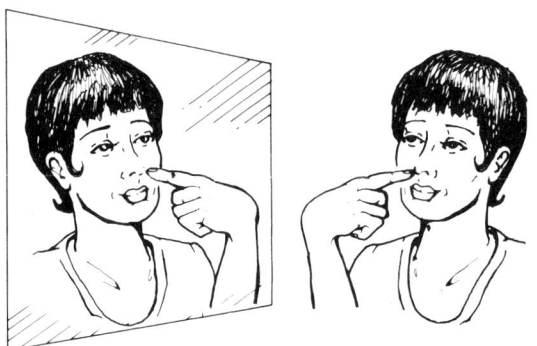

The **image** appears to be the same distance behind the mirror as the object is in front. Touch your nose with your **right** hand and your image touches its nose with the left hand. Left and right are reversed for an object and its image.

MIXED NUMBER

A **number** consisting of an **integer** and a **proper fraction**.
Examples: $3\frac{1}{4}$, $2\frac{7}{8}$, $-8\frac{2}{3}$.

MÖBIUS STRIP

A **surface** with only one side. It can be made by turning one end of a long strip of paper through $180°$ and fastening it to the opposite end.

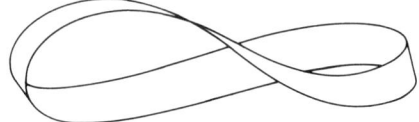

A mobius strip has only one **edge**. It is named after a German August Möbius (also spelt Moebius).

MODE

One type of **average** or **mean**. It is the **value** that occurs most frequently.
Example: $2, 2, 3, 3, 3, 3, 4, 4, 5, 5, 5.$
3 is the mode.

MODERN MATHEMATICS

There are many different interpretations:

1 A logical system of **mathematics** developed from **statements** which are accepted without **proof** (**axioms**). From these **theorems** are deduced and a system built up.

2 The collection of new topics that have been introduced in recent **years**. These include **topology**, **set theory**, linear programming and **computer** studies but there are many others.

3 In the primary school it includes new topics (**tessellations**, **base** work, sets, ...) and also new methods of teaching old topics (projects, discovery, environmental mathematics, ...).

MODULAR ARITHMETIC

Also called modulo or **clock arithmetic**.
This is best explained by means of an example:

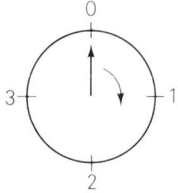

We will use modulus 4. The clock **face** shows only 0, 1, 2 and 3. Starting at 0 and turning through 2 **right angles**, then 3, the hand would point to 1. We write $2+3=1$ (mod 4). Similarly $3+1=0$ and so on for other **values** as shown on this **table**:

+	0	1	2	3
0	0	1	2	3
1	1	2	3	0
2	2	3	0	1
3	3	0	1	2

Subtraction, **multiplication** and **division** can also be carried out in any modulus. Mod 5 uses 0, 1, 2, 3 and 4. The highest **digit** involved is always one **less than** the modular number. Mod 2 requires 0 and 1. For mod n we have 0, 1, 2, ...(n−1). The clock face may help but is not essential. A study of modular arithmetic helps us to see the **structure** of **mathematics**, particularly the following **properties**: **commutative**, **associative**, **distributive**, **closure**, **identity** and **inverse**.

MONTESSORI APPARATUS

A wide variety of materials for primary children as aids for learning **mathematics**. It was developed by Dr. Maria Montessori (1870–1952). She was one of the first educators to stress the importance of children using concrete materials such as bricks, **rods**, cut-out **shapes**, etc.

Dr. Montessori used rods that have now been developed into the **Stern**, **Cuisenaire** and **Colour Factor** apparatus. An emphasis was placed on using all the senses when learning and not only sight.

MONTH

Approximately one twelfth of a **year**.
There are various meanings:

1 The **calendar** month has 28, 29, 30 or 31 days.

2 Any period of 4 **weeks** or 28 days.

3 The lunar month is the time between recurring phases of the moon. It is approximately $29\frac{1}{2}$ days.
There are also sidereal, solar, synodic and periodic months.

OLD ENGLISH *monath*, month; from *mona*, moon.

MORE THAN
(see GREATER THAN >)

More correctly 'greater than'. 8 is more than 5. This is written $8 > 5$. Is **less than** is written as $<$. For instance $5 < 8$.

MOTION GEOMETRY

An approach to **geometry** via movement.
Example:

1 The **line segment** AB, moved in the **direction** shown ($\rightarrow$), sweeps out or generates the **parallelogram** ABCD. (See **generator**.)

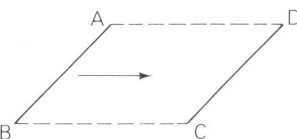

2 A **right-angled triangle** rotated about AB will generate a **cone**.

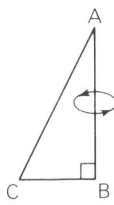

3 The **properties** of a **rectangle** can be investigated by fitting it into an outline.

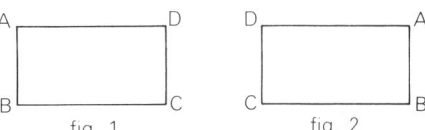

fig. 1   fig. 2

Draw round a rectangle ABCD. Turn the rectangle over to get Fig. 2. AB is seen to be **equal** to DC, **angle** A = angle D, angle B = angle C.

LATIN *motio*, to move; *ge*, Earth; *metron*, **measure**.

MULTIBASE ARITHMETIC BLOCKS (M.A.B.)

This material is used in the teaching of **place value** and ideas that use various **number bases**.

Four of the **units** shown below make one long. Four longs (or sixteen units) make one **flat**. Four flats (or sixteen longs, or sixty-four units) make one **block**.
Example:

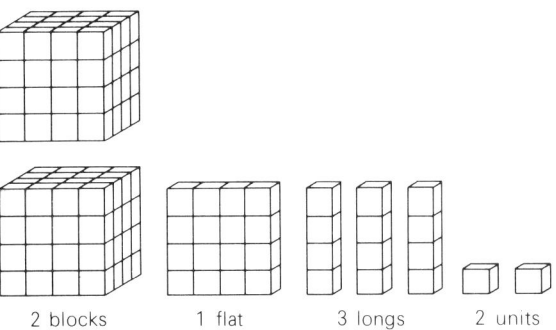

| 2 blocks | 1 flat | 3 longs | 2 units |

This is recorded in base four as 2132_{four}. We can change the **amount** from base four to base ten as follows: $2132_{four} = 2(64) + 1(16) + 3(4) + 2 = 128 + 16 + 12 + 2 = 158$. This could be written as 158_{ten} but it is usual to assume a number is in base ten, unless stated otherwise, and just write 158.

MULTIPLE

Any **number** that has a given **whole number** as a **factor** is called a multiple of that factor.
Examples: 48, 24 and 18 have 3 as a factor so they are all multiples of 3. They are also multiples of 2 and 6.
$8 \times 5 = 40$. 40 is a multiple of 5 and also of 8.

MULTIPLICAND

The **number** or **quantity** that is to be multiplied.
Example:

$$\begin{array}{r} 24 \quad \textbf{(Multiplicand)} \\ \times 3 \quad \textbf{(Multiplier)} \\ \hline 72 \quad \textbf{(Product)} \end{array}$$

MULTIPLICATION

The process of repeated **addition**.
$3 + 3 + 3 + 3 = 4 \times 3 = 12$.
In the **terms** of **modern mathematics** multiplication is a **binary operation** (an operation on two **elements**). In the above example 4 and 3 are the elements.

MULTIPLIER

The **number** by which another is multiplied.
Example:

$$\begin{array}{r} 14 \quad \textbf{(Multiplicand)} \\ \times 8 \quad \text{Multiplier} \\ \hline 112 \quad \textbf{(Product)} \end{array}$$

MULTIPLY

See MULTIPLICATION.
To carry out the process of multiplication.

N

NAIL BOARD
See GEO-BOARD.

NAPIER, JOHN 1550–1617
A Scottish mathematician who invented **logarithms** and a system of **rods** for **multiplication** known as **Napier's Bones**.

NAPIER'S BONES
A device for multiplying named after their inventor John Napier. Ten strips, headed 0 to 9, carry the **multiplication tables** and another strip acts as an index.
In some forms the headings 0 to 9 are omitted.

To find 63×5. Place the 6 and 3 rods **side** by side. Against 5 we have

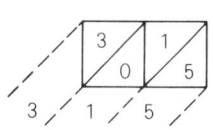

index

	6	3
1	6	3
2	1/2	0/6
3	1/8	0/9
4	2/4	1/2
5	3/0	1/5
6	3/6	1/8
7	4/2	2/1
8	4/8	2/4
9	5/4	2/7

(1)

Add along the **diagonal** to get the answer, 315. Napier made the first ones from bone.
OLD ENGLISH *ban,* bone.

NATURAL NUMBERS
The **numbers** 1, 2, 3, 4, 5, 6, 7, … Also called **counting numbers** or **positive integers**.

NAUGHT
See NOUGHT.

NAUTICAL MILE
Also called geographic **mile**.

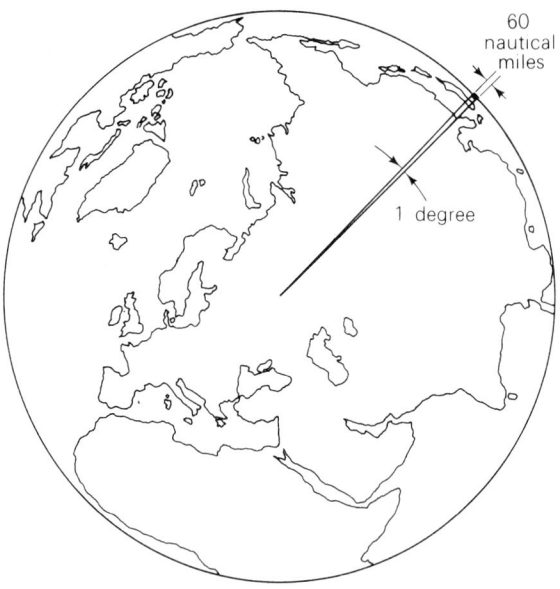

The **diagram** shows $\frac{1}{360}$ of a **great circle**, that is 1 **degree**. A degree is divided into 60 **equal** parts and each is called a **minute**. The **length** of the small **arc** that this **angle** makes on a great circle is called a nautical mile.

It follows that 60 nautical miles equal 1 degree (degree is here used as a **measure** of distance). As the Earth is not a perfect **sphere** the length would vary from place to place. An **average value** is therefore used and the nautical mile is taken as 6080 **feet** (1.85 **kilometres**).

GREEK *nautilos*, a sailor.
LATIN *mille*, a **thousand**.
(A Roman mile was a thousand **paces**.)

NAVIGATE
To direct or plan the course of a ship or aeroplane. This may involve using such aids as a map, the stars, **compass**, radar or wireless.

NEGATIVE NUMBERS
Numbers less than **zero**.
Example: ⁻2 (also written as −2 but then likely to be confused with the **operation** of **subtraction**).
Negative numbers can be represented on a **number line**. They are to the left of zero.

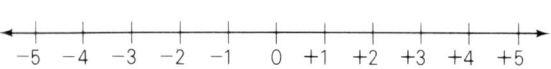

Two number lines at **right angles** form the axes of a **coordinate** system.

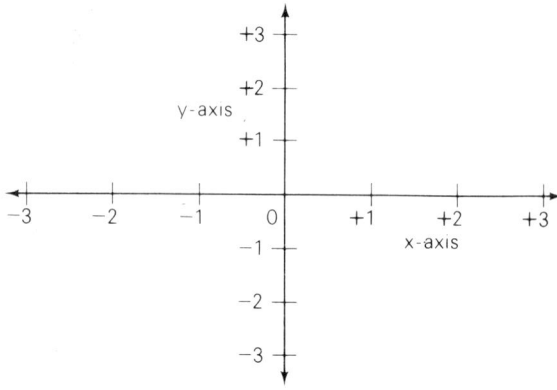

NET
1 A **plane shape** that when folded forms a **solid**.
Examples:

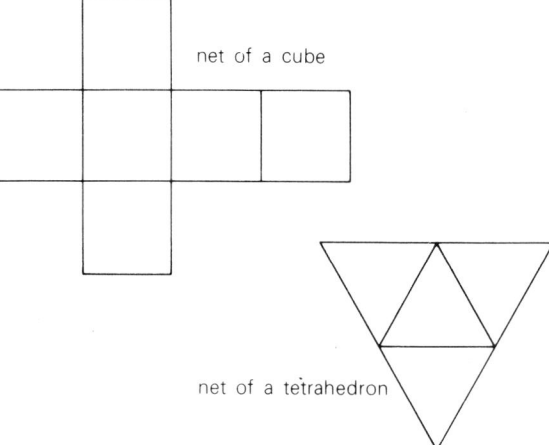

net of a cube

net of a tetrahedron

(A solid does not have to have material inside it. A balloon is, in **mathematics**, a solid.)

2 Net **weight** (sometimes spelt nett). The weight of goods after subtracting the weight of the container and packing, (see **gross weight**).

3 Net price (sometimes spelt nett). The price after deducting for postage, packing and other expenses.

4 Net (or network). **diagram** showing **points** (called **nodes** or junctions) and **arcs**.

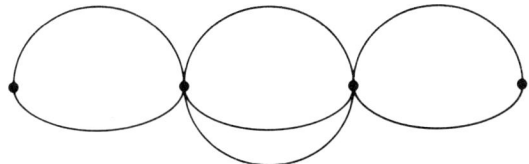

Their **properties** are studied as one section of **topology**.

NEWTON, 1642–1727

1 Sir Isaac Newton, was an English mathematician and scientist. He invented the **calculus** and established the laws of motion and **gravity**. At school he was inattentive and near the bottom of his class yet he became one of the greatest scientists the world has known.

2 A **unit** of **force** in the **S.I. system**. It is the force required to produce an **acceleration** of one **metre** per **second** per second in a **mass** of one **kilogram**.

NODE

A **point** where **lines** or **curves intersect**. Also called a junction.

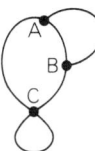

A, B and C are nodes.

NOMOGRAM

Also called nomograph. A **diagram** designed to give a **value** to a **variable**, when other related ones are known. There are generally three variables shown on **straight** or **curved lines**. Given any two of these the third can be found. *Example:*

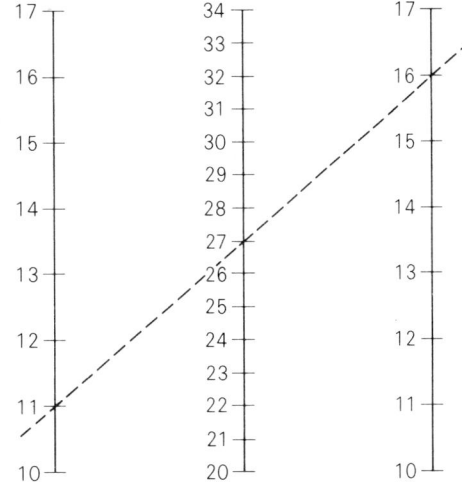

The line shows how the nomogram can be used for **addition** or **subtraction**. The edge of a ruler can be placed so as to join the two **numbers** you wish to add (11 and 16) or subtract (27 − 11). The example above could be for $11 + 16 = 27$, $27 - 11 = 16$ or $27 - 16 = 11$.

NONAGON

Strictly speaking a **polygon** with nine **angles** but it is more often thought of as one with nine **sides**.

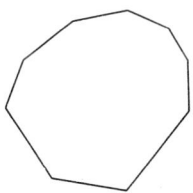

an irregular nonagon

If the sides are **equal** in **length** and the angles are all equal then the polygon is **regular**. It is therefore a regular nonagon.

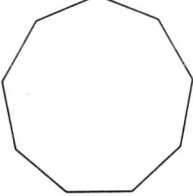

a regular nonagon

LATIN *nonus*, ninth.
GREEK *gonia*, angle.

NONE

See **zero** and **nothing**. Not any. Literally 'not one'.

NORM

A **term** used in **statistics**, meaning the **normal** or **average**.

NORMAL

1 At **right angles** or **perpendicular**.
Example:

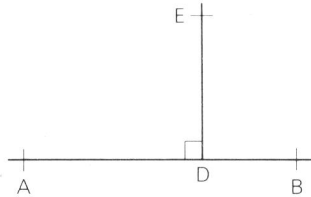

ED is drawn normal to AB.
The **term** is also used for the **line** itself.
Example: In the above **diagram** ED is the normal to AB at D.
The normal to a **curve** at a given **point** is the line through that point at right angles to the **tangent**.

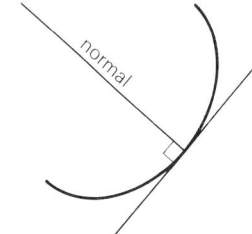

2 See NORM.

NORMAL CURVE OF DISTRIBUTION

A bell shaped **curve** that frequently occurs in **statistics**.

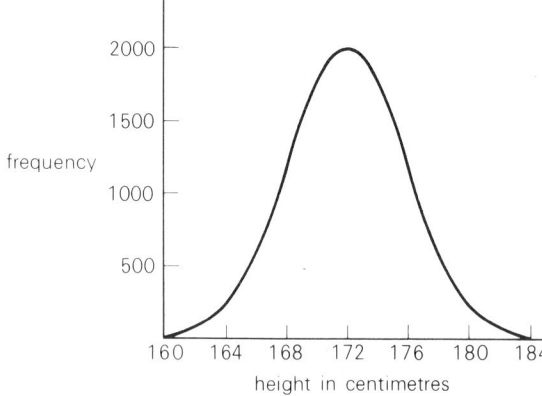

Example: A **graph** of the heights of a large **number** of people is found to have many around the middle **region** but few who are very tall or very short. Such graphs are approximately symmetric.

NORTH

1 Magnetic North. The **direction** in which the **compass** points.

2 Geographic or True North. The direction of the North Pole. The North Pole is at the north end of the **axis** on which the earth **rotates**.
The Magnetic and geographic North are not the same. The **difference** between them changes gradually from **year** to year and from one place to another.

NOTATION

A system of **symbols** representing **numbers**, **operations** or any other mathematical ideas.
Example: 0, 1, 2, 3, 4, 5, 6, 7, 8 and 9 are symbols and so are $+, -, \times, \div, \sqrt{}$. (The last five are also called **signs**.)
Other symbols can be seen in the following entries: **Roman system, Egyptian system, Babylonian system.**

NOT EQUAL

Not the same. **Unequal**.
5 is not **equal** to $3+1$. This is written as $5 \neq 3+1$.

NOTHING

See **zero** which is a more correct **term**.
None. The absence of any **number**, **quantity** or **amount**. Literally 'no thing'.

NOUGHT

Also **naught**.
Zero 0.

NULL SET

Empty set. A **set** with no **number**. It is written as $\{\ \}$ or $\varnothing$.

NUMBER

A **measure** of **quantity** but also used in an abstract way without relating to 'how many' or **measurement**. There are many kinds of numbers:

1 **Natural numbers** (1, 2, 3, 4, ...)

2 **Whole numbers** (0, 1, 2, 3, 4, ...)

3 **Integers** (... ⁻4, ⁻3, ⁻2, ⁻1, 0, ⁺1, ⁺2, ⁺3, ⁺4, ...)

4 **Rational numbers.** This includes **fractions** and all the numbers in 1, 2 and 3 above.
There are also **irrational, real, complex, square, triangular, polygonal, prime** and many other kinds of number.
(See CARDINAL NUMBER and ORDINAL NUMBER.)

NUMBER LINE

A **line** in which **points** correspond to **integers** and intermediate **positions** correspond to other **real numbers**.

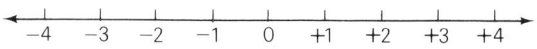

Counting, **addition**, **subtraction**, **multiplication** and **division** can be illustrated by means of the number line.

NUMBER SENTENCE

A mathematical **sentence**.

Examples: $3 + 2 = 5$

$9 > 6$

$\square + 3 = 10$

NUMERAL

Any **symbol** or name for a **number**.

Examples: $8, 23, 114, \frac{8}{11}$.

Seven, twenty-two.

IV, XI.

NUMERATOR

That part of a **fraction** which is in the top.

$3 \leftarrow$ numerator

$\overline{5} \leftarrow$ **denominator**

It shows 'how many' of the **equal** parts there are. In $\frac{6}{11}$ the 6 shows there are six of the $\frac{1}{11}$ths. ($\frac{6}{11}$ or $\frac{1}{11} \times 6 = 6 \times \frac{1}{11}$). $\frac{6}{11}$ could also be interpreted as 6 separated into 11 equal **amounts**.

NUMERICAL

1 Relating to or denoting **number**.

2 Numerical **value**. The value when the **sign** or **direction** is ignored.

Example: -3 or $^-3$ has a numerical value of 3.

O

O

1 A label for a **point** especially the **centre** of a **circle** or the **origin** of **coordinate axes**.

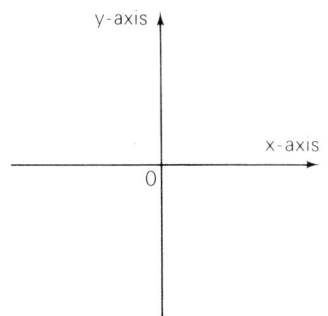

2 **Zero.**

3 Roman **numeral** in the Middle Ages.
0 represented 11 000.
Derived from the Greek letter omicron.

OBLATE

Flattened at one or more **points**. Especially used when two points are flattened that are opposite one another. The Earth is oblate as it is slightly flattened at both poles.

OBLIQUE

Slanting. An oblique **line** is neither **parallel** nor **perpendicular** to a given **direction**.
Example:

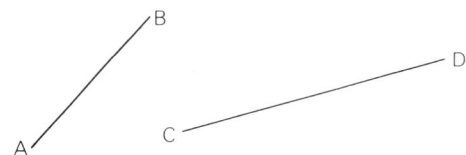

AB is oblique with reference to CD. Similarly CD is oblique to AB.

OBLONG

A **rectangle** with adjacent **sides** that are **unequal**.
A **square** is a rectangle but it is not an oblong because the adjacent sides are **equal**.

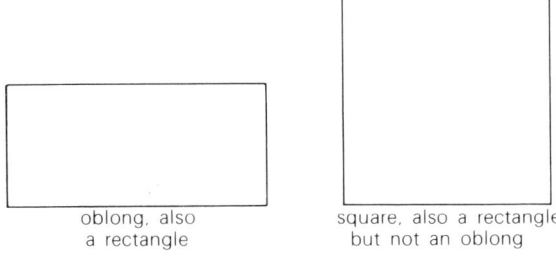

oblong, also a rectangle

square, also a rectangle but not an oblong

OBTUSE ANGLE

An **angle greater than** 90° but **less than** 180°.

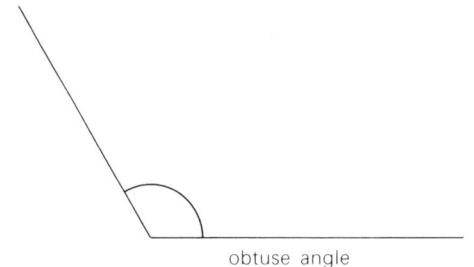

obtuse angle

(See ACUTE ANGLE.)
LATIN *obtusus*, blunt.

OCTAGON

A **polygon** with eight **angles** and therefore with eight **sides**.

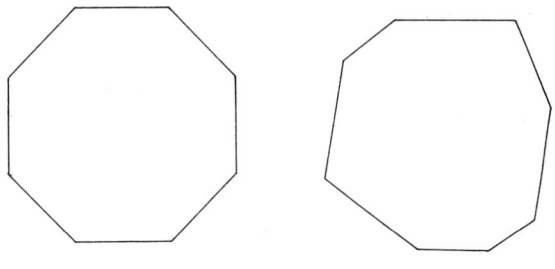

a regular octagon   an irregular octagon

A **regular** octagon has eight **equal** angles (135°) and eight equal sides. An octagon that is not regular is irregular.
GREEK *okta* (and okto), eight; *gonia*, an angle.

OCTAHEDRON

A **solid** with eight **faces**.

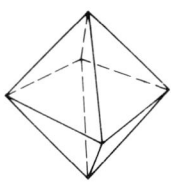

a regular
octahedron

GREEK *okta*, eight; *hedra*, seat (face).

ODD NUMBER

A **whole number** which has a **remainder** of 1 when divided by 2. A whole number which is not **even**.

ODDS

The **ratio** of the **probability** that an event will happen to the probability that it will not.
Example: What are the odds in favour of getting a 1 or 2 when rolling a **die**? There are two events in favour (1, 2) and four against (3, 4, 5, 6). Odds in favour $= \frac{2}{4} = \frac{1}{2}$. Similarly the odds against are $\frac{2}{1}$ or 2 to 1 against.

ODOMETER

Also called a hodometer.
An **instrument** used to **measure** the distance travelled. It is normally attached to a wheel.
GREEK *hodos*, a way; *metron*, a measure.

OHM

A **unit** of electrical resistance in the **S.I. System**. Named after the German electrician Georg Ohm (1787–1854).

ONE DIMENSION

A **figure** having only **length**.
Example: A **line** has length only. A **rectangle** has length and **breadth** and therefore has **two dimensions**.

ONE-TO-ONE CORRESPONDENCE

A **relation** between two **sets** such that (i) every **member** (or **element**) in one set **matches** with one, and only one member of the second set. Also (ii) every member of the second set matches with one, and only one, member of the first set.

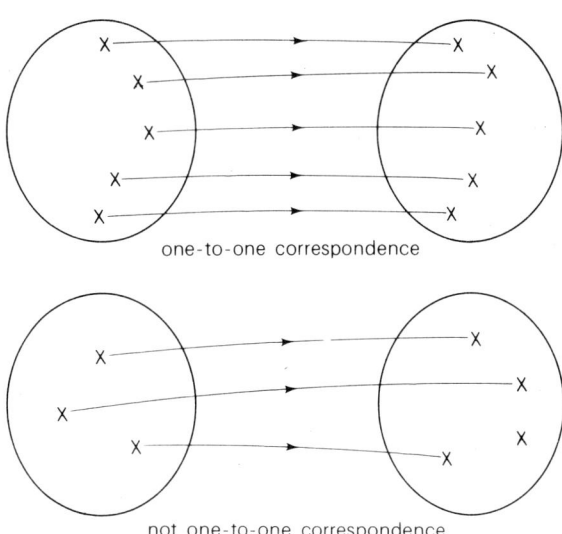

one-to-one correspondence

not one-to-one correspondence

(See ARROW LINE and map.)

OPEN CURVE

A **continuous curve** whose end **points** do not **meet**.

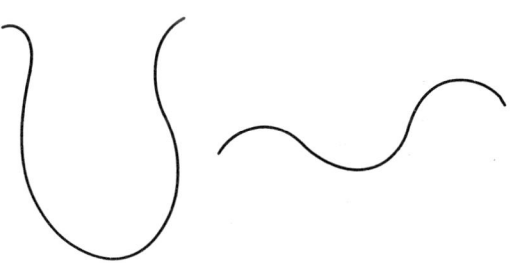

These are open curves.

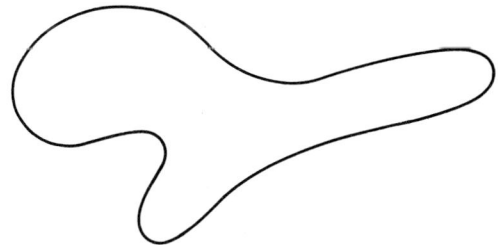

This is a **closed curve**. The end points meet.

OPEN SENTENCE

A mathematical **sentence** that **includes** a **variable**. For instance $2x + 1 = 17$. x is the variable and the sentence becomes a **true statement** if x is replaced by 8. For any other **value** of x the statement would not be true.

Example: $3 + \square = 10$ is an open sentence. It becomes a true statement if the **place holder** $\square$ stands for 7. For any other value it would be a **false sentence** or statement.

OPERATION

A means of combining **numbers**, **sets** or other mathematical **elements**.

Addition, **subtraction**, **multiplication** and **division** are well known operations on numbers. **Union** and **intersection** are operations on sets.

OPPOSITE ANGLES   See **vertically opposite angles**.

OR

An important word in **logic**. It is represented by the **symbol** $\vee$. It is used to connect clauses as in 'I will go for a walk or watch television'. It is therefore called a connective.

If A **represents** 'I will go for a walk' and B represents 'I will watch television' the **sentence**, in logical symbols, is written as A $\vee$ B. Or is a shortened form of other.

ORBIT

The path in **space** of a body, especially a planet, star or satellite. Many orbits are **ellipses**.

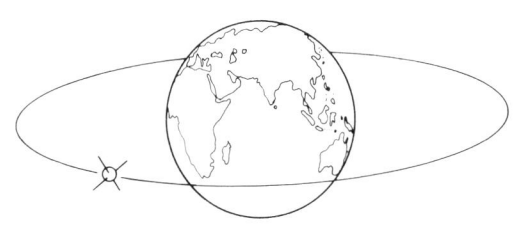

ORDER

1 An arrangement according to **size**, **quantity** or **value**.
Example: 1, 3, 8, 12, 32 are in **ascending order** of **magnitude**. 198 cm, 1 m, 23 cm are in **descending order** of length.

2 Order of an **equation**. The highest **power** or **degree** of a variable.
Example: $x^3 + 3x^2 + x - 7$ is of the third order (or degree).

3 Order of **symmetry**. If on rotating a **figure** fits n times into its own outline it has **rotational symmetry** of order n. *Examples:*

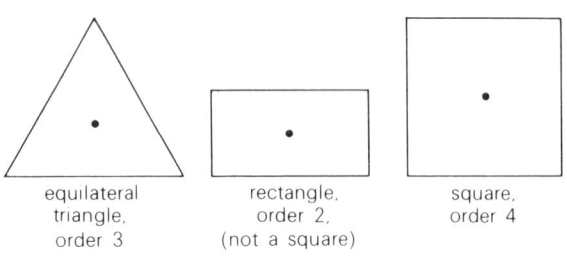

equilateral triangle, order 3 — rectangle, order 2, (not a square) — square, order 4

ORDERED PAIR

A **pair** of **numbers** or **symbols** in which the **order** is important. Using **Cartesian coordinates** the **point** (3, 2) is different from the point (2, 3).

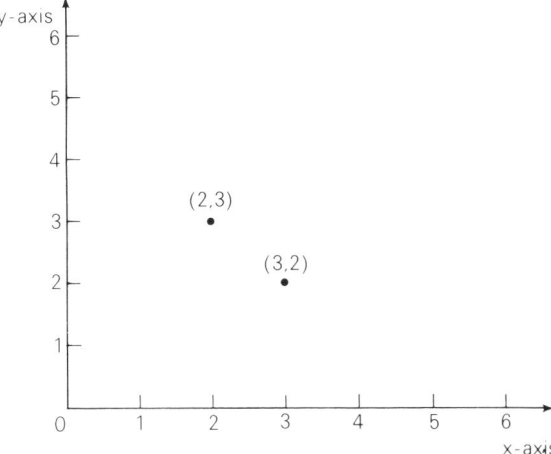

In a **relation** such as 'is taller than' the ordered pair (John, Mary) indicates that John is taller than Mary. The order is important as it is not correct to write (Mary, John). This would mean 'Mary is taller than John'.

ORDINAL NUMBER

The **number** denoting the place when **elements** are arranged in **order** according to some **property** such as **length**, **volume** or some other **measure**.
Example: 3rd in a race. (The ordinal number is 3.) 8th of January. (The ordinal number is 8.)

ORDINATE

In **graph** work the second of an **ordered pair** denoting a **point**. In (3, 5) the 5 is the ordinate (or y **value**).

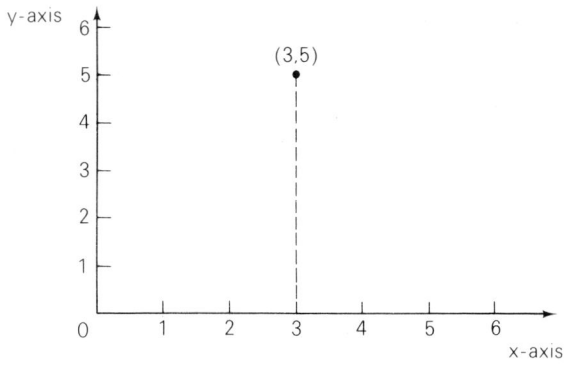

The ordinate is the distance from the **horizontal axis** x-**axis**).

ORIENTATE

To place in a certain **direction** with **relation** to the **points** of the **compass** or any other fixed direction.

Example: A ship is sailing on a **bearing** of 170°. It is orientated with respect to magnetic **North** by means of a compass.

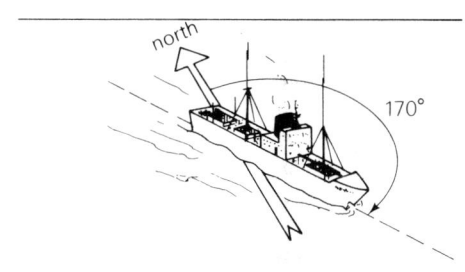

ORIGIN

A fixed **point** from which **measurements** are taken. The point where *x* and *y* **axes** meet, indicated by 0. See **Cartesian coordinates**.

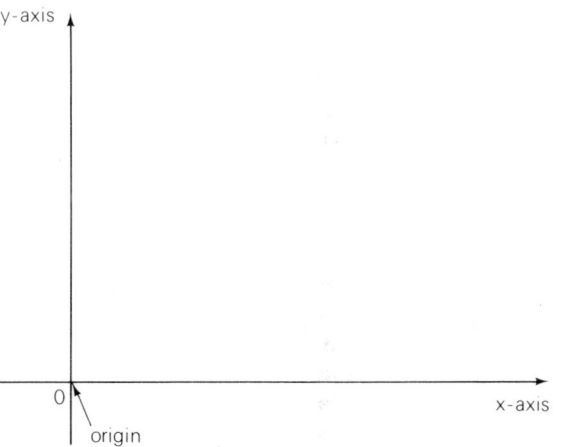

OSCILLATE

To swing to and fro. Vibrate.
Example: When a **pendulum** is in motion it is said to oscillate.

OUNCE

A **unit** of **mass** or **weight** in the **imperial system** of **measurement.** (avoirdupois)
16 ounces = 1 **pound** (16 oz = 1 lb).
1 ounce is approximately 28.35 **grams**.
LATIN *uncia*, the twelfth part. So called as in the troy **measure** there were twelve ounces in the pound. The word was retained for the avoirdupois measure even though it had 16 ounces in a pound.

OVAL

This word is used in two different senses:
1 A **two dimensional figure** that is egg shaped. One end is more pointed than the other. It is **symmetrical** about the **axis** shown.

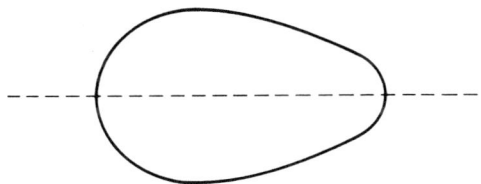

2 An **ellipse**. The ends are the same **shape** and the **figure** is symmetrical about the two axes shown.

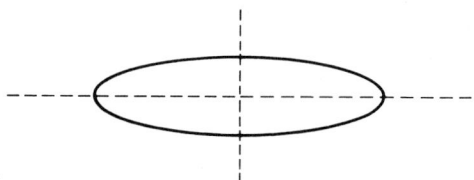

LATIN *ovum*, egg.

OVOID

An egg-shaped **solid**. The **term** is sometimes used for a **plane figure**.
(See OVAL.)
LATIN *ovum*, egg.

P

p
Abbreviation for **penny** or pence. p was introduced in 1971 when our money changed to **a decimal** system.

PACE
1 The distance between the feet when walking, measured from heel to heel.

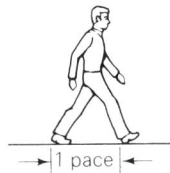

The Roman pace was a **double** one and 1000 such paces equalled the Roman **mile** (mille, in Latin means a 1000).

2 The **rate** at which something is done.
Example: He walked at a **fast** pace.

PAIR
1 Two.
Example: The children formed in pairs for the dance.

2 See ORDERED PAIR.

PALM
The **width** of the hand.
1 palm is approximately 4 **digits**.

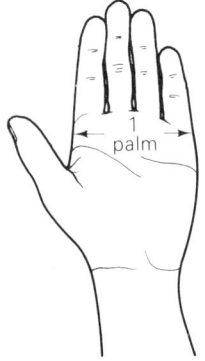

PANTOGRAPH
An **instrument** for enlarging drawings. See picture opposite.
GREEK *pan*, everything; *graphein*, to write.

PAPER FOLDING
1 An art known as Origami. Many interesting **patterns** and objects can be made.

2 A means of illustrating mathematical **properties** by folding paper.
a A **straight line** can be made.

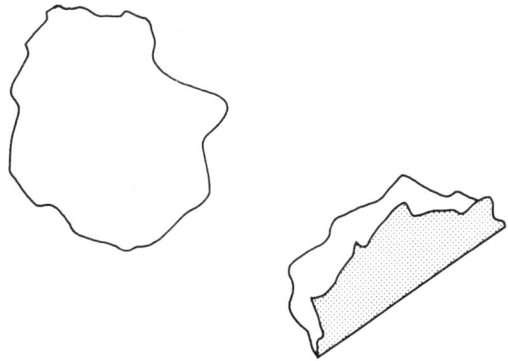

b By folding again we have a **right angle**.

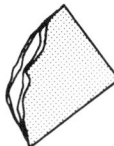

Some other possibilities: **Angles** can be bisected, **squares** made, **symmetrical shapes** constructed, lines bisected.

PARABOLA
If a **plane** cuts a **cone parallel** to the **line** AB (that is a line from the **vertex** A to any **point** on the **perimeter** of the **base**)

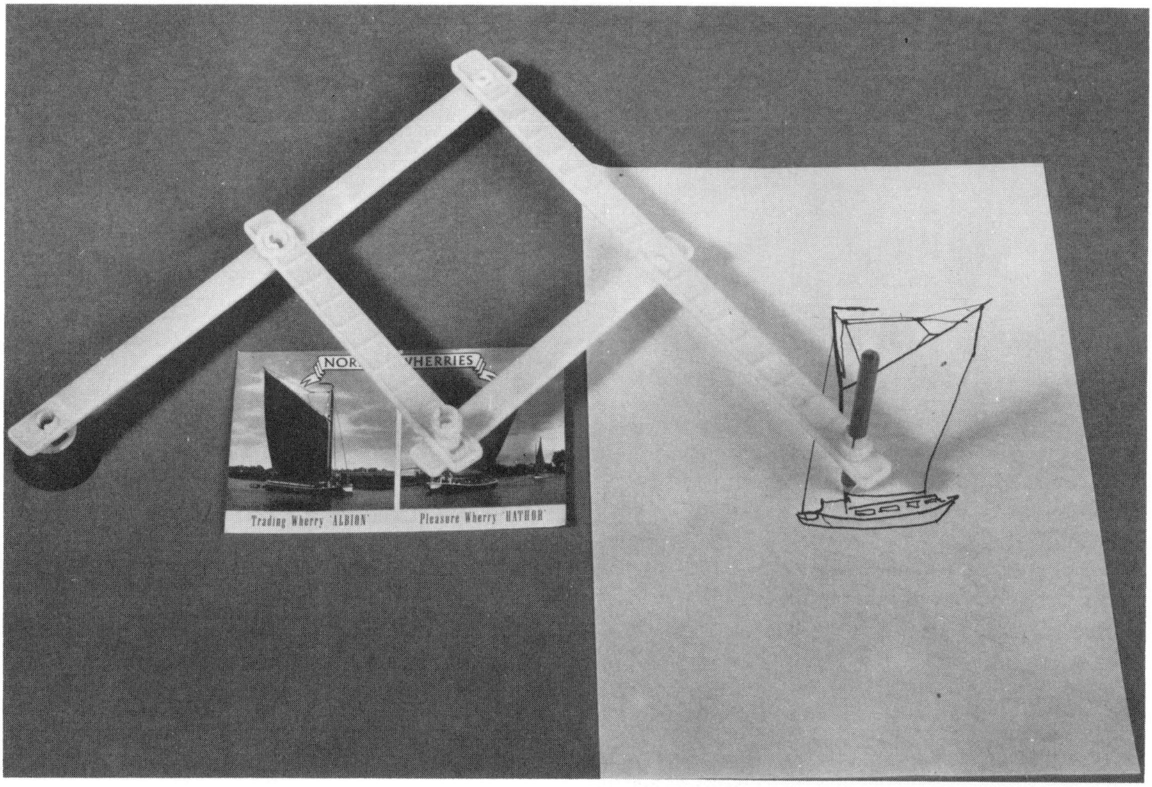

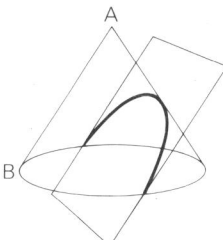

a parabola is formed. A parabola can also be described as the **set** of points such that the distance from any one of them to a fixed line is **equal** to the distance to a fixed point (called the **focus**).

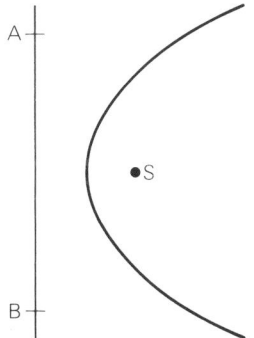

S is the focus, or fixed point.
AB is the fixed line or directrix.

Parabolas formed by water in a fountain at Steine Gardens, Brighton.

PARADOX

A **statement** in which two conclusions appear to contradict each other and yet both seem to be true.

Example:

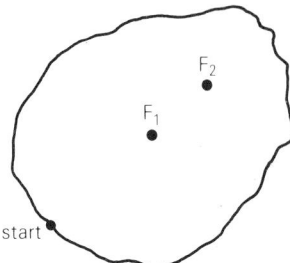

A frog jumps half way across a pond at his first jump (to F_1). He then jumps half of the remaining distance (to F_2). Next he jumps half the distance still remaining and so on. It appears that as the frog keeps on jumping he must eventually reach the other **side**. It also could be argued that whatever the distance of the last jump that same distance still remains, so he never reaches the other side. We have a paradox.

We can in fact prove that the frog would never reach the other side of the pond.

If we call the distance across the pond 1 **unit** the **length** of each jump is

$\frac{1}{2}, \frac{1}{4}, \frac{1}{8}, \frac{1}{16}, \frac{1}{32}, \frac{1}{64}, \ldots$ No matter how many **terms** we **add** the **sum** is never quite 1 unit.

$$\frac{1}{2} + \frac{1}{4} + \frac{1}{8} + \frac{1}{16} + \frac{1}{32} + \frac{1}{64} \cdots$$

Still to cover $\quad \frac{1}{2} \quad \frac{1}{4} \quad \frac{1}{8} \quad \frac{1}{16} \quad \frac{1}{32} \quad \frac{1}{64} \quad - - -$

GREEK *para*, beyond; *doxa*, opinion.

PARALLEL

Having no **points** in common no matter how far the **lines** or **surfaces** are produced.

1 Parallel lines

The arrows show the lines are parallel.

2 A line parallel to a **plane**.

3 Two planes that are parallel.

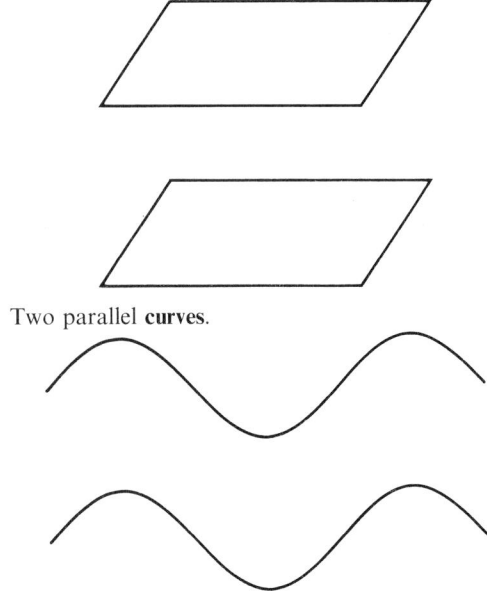

4 Two parallel **curves**.

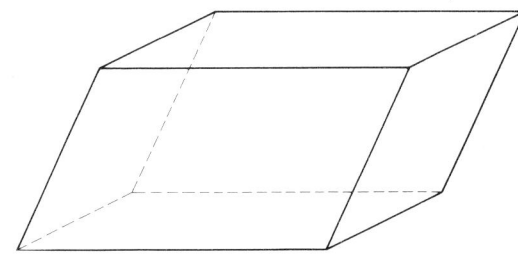

Also see LATITUDE for parallel of latitude.

PARALLELEPIPED

Also spelt parallelopided but this is not strictly correct. A **solid** with six **faces**, each **pair** of opposite faces being **equal parallelograms**.

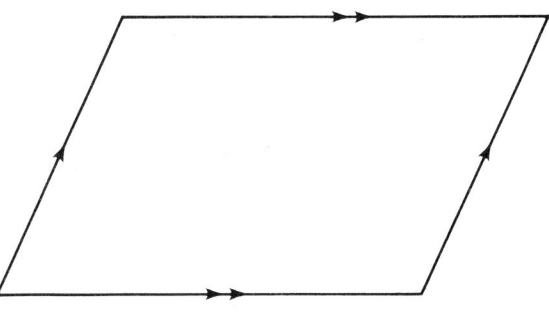

If the **angle** between every pair of **planes** that meet is 90° the parallelepiped is a **cuboid**. If it is not 90° the **figure** is called an **oblique** parallelepiped.

PARALLELOGRAM

A **quadrilateral** with both **pairs** of opposite **sides parallel**.

The arrows show which **lines** are parallel.

PARENTHESES

The **symbol** ( ) enclosing **terms** that are to be grouped together.

Example: $2+(4 \times 5) = 2+20 = 22.$
but $(2+4) \times 5 = 6 \times 5 = 30.$

Also called **brackets**, a term which includes [ ] and { } **(braces)**. Thus all brackets are not parentheses but parentheses are one form of brackets.

PARTIAL PRODUCT

The **product** resulting from **multiplication** by a **number** when this forms only one part of the **computation**.

Example.

$$\begin{array}{r} 38 \\ \times 24 \\ \hline 760 \\ 152 \\ \hline 912 \\ \hline \end{array}$$

$760 \leftarrow$ Partial product (From 38×20)
$152 \leftarrow$ Partial product (From 38×4)

PARTITION

1 One aspect of **division**, the other being **quotition**. Partition, also called **sharing**: $36 \div 4$ is thought of as sharing 36 objects between 4 people. (Quotition is 'How many 4's are there in 36?')

2 To separate a **set** into **subsets**.

Example: The set of **natural numbers less than** 10 can be partitioned into subsets of **odd** and **even numbers**.
Odd numbers $= \{1, 3, 5, 7, 9\}$. Even numbers $= \{2, 4, 6, 8\}$.

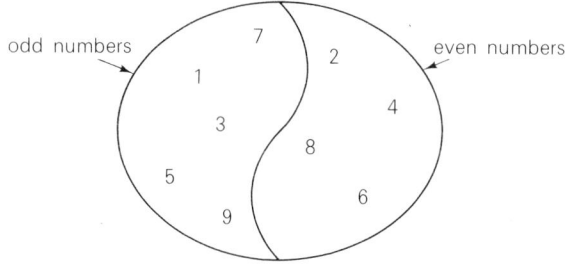

PASCAL'S TRIANGLE

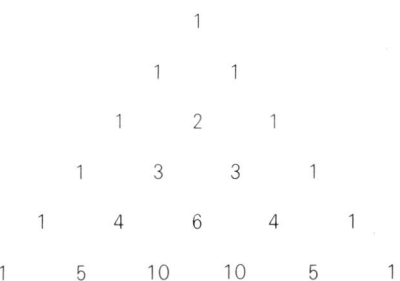

More **rows** could be added in this way: Start and finish with 1. The other **numbers** are found by adding the two

nearest ones from the row above. The **triangle** is named after a French mathematician philosopher and scientist, Blaise Pascal (1623–1662). If the numbers in each row are added the **sums** are 1, 2, 4, 8, 16, 32, and so on, doubling each time. Pascal's triangle has many uses in advanced **mathematics**.

You will find some interesting **patterns** in the numbers along **diagonals**.

PATTERN

A systematic arrangement of **numbers**, **shapes** or other **elements** according to some **rule**.

Examples:

a 3, 5, 7, 9, 11   (adding 2)

b 1, 2, 4, 7, 11, 16, 22   **add** 1, then 2, then 3, increasing the number added by 1 each **time**.

c

PECK

An **imperial unit** of **capacity** for dry goods.
1 peck = 8 **quarts** or 16 **pints**.

PEDOMETER

An **instrument** for recording the **number** of steps taken. From this the distance walked can be calculated.

LATIN *pedis*, **foot**.
GREEK *metron*, **measure**.

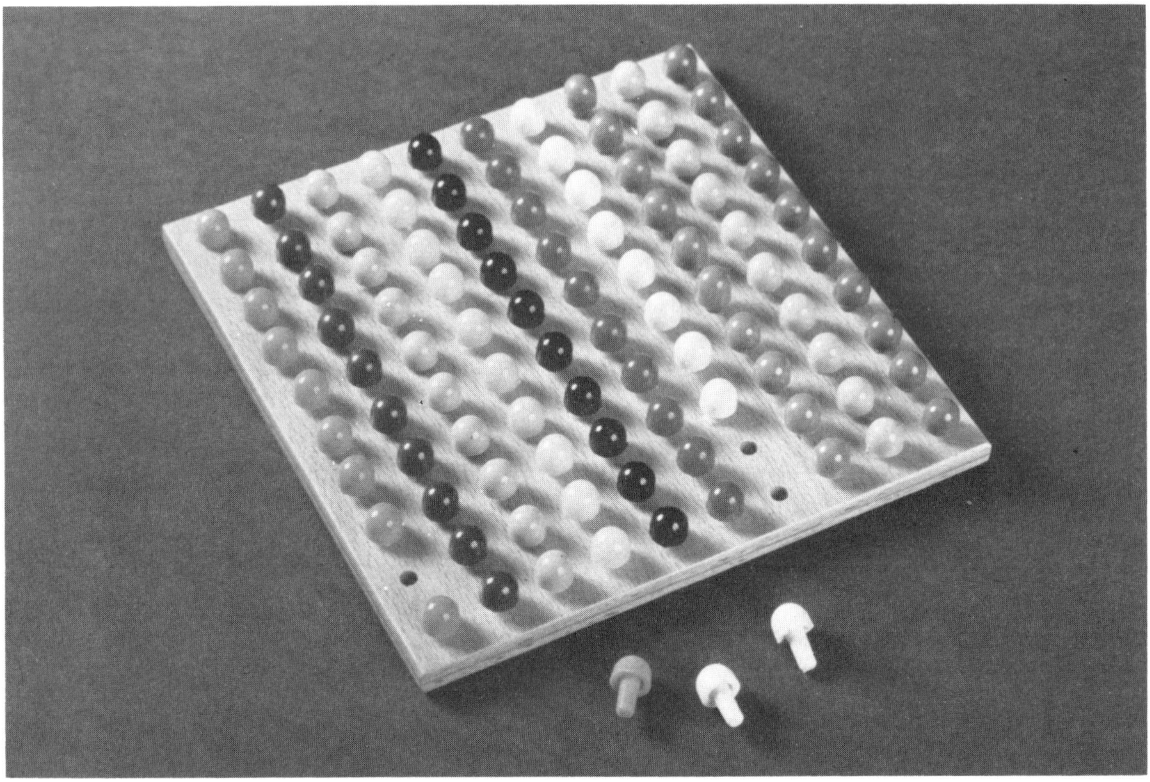

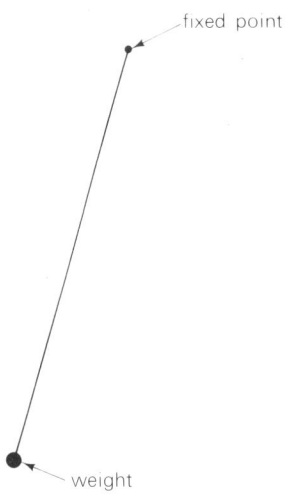

PEGBOARD

A board with holes, generally arranged in **rows** and **columns**. Coloured pegs can be used to show **number** bonds and **patterns**.

PENDULUM

1 Simple pendulum. A small **mass** or **weight** is suspended from a fixed **point**. The mass can be considered to be concentrated at one point – the **centre** of **gravity**.

2 Compound pendulum.

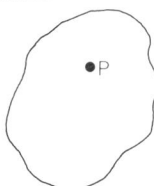

Any body that **oscillates** (or swings) about an **axis** (The axis is through P in the **diagram**).

3 **Seconds** pendulum. A pendulum that takes one second to swing from one extreme **position** to the other, that is from A to B in the diagram.

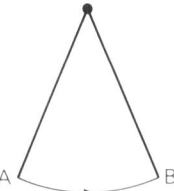

A seconds pendulum is about 1 **metre** in **length**. This differs slightly from one place to another. In 1581 **Galileo** saw a lamp swinging in Pisa cathedral and noticed that each swing took the same **time**. This led to his study of the pendulum.

LATIN *pendulus*, hanging.

PENNY

A coin.
One hundred are the same **value** as one **pound**. 100p = £1.

(See DENARIUS.)

PENTAGON

A **polygon** with five **sides**.
If the sides are all the same **length** and the **angles** all **equal** then the pentagon is **regular**.

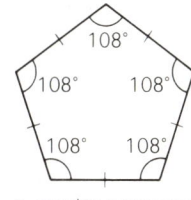

a regular pentagon

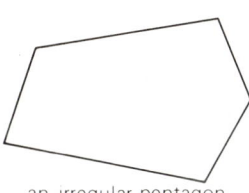

an irregular pentagon

GREEK *pente*, five; *gonia*, angle.

PENTAGRAM

A five pointed star.
The pentagram of **Pythagoras** is obtained by joining the vertices of a **pentagon**.

This pentagram was the secret **sign** of the Society of Pythagoras.

GREEK *pente*, five; *gramma*, a letter.

PENTOMINOES

A **polyomino** with five **squares** joined along their **sides**.

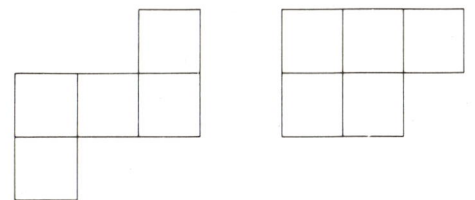

There are twelve different pentominoes. The **shapes** are cut out of squared paper or card. Two pentominoes are regarded as the same if, when rotated or turned over, they fit on to one another.

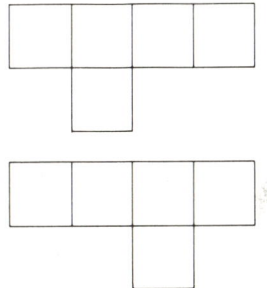

These two pentominoes are the same.

GREEK *pente*, five.
The 'omino' comes from the second part of **domino**.

PER CENT (OR PERCENT), PERCENTAGE

A means of representing a **rate** or **ratio** between two **amounts**. One is expressed as a **number** of hundredths of the other. 25 per cent is written as 25% and is **equivalent** to $\frac{25}{100}$, $\frac{1}{4}$ or 0.25.

Example: Find 5 per cent of 60. $\dfrac{5}{100} \times 60 = 3$. The

The 3 is called the percentage. Note that percentage is a number whereas per cent is a rate.

LATIN *per*, for each; *centum*, a hundred.

PERCENTAGE ERROR

The error expressed as a **percentage**.
Example: 1 centimetre error in measuring 1 **metre** is $\frac{1}{100}$ as a **fraction**, or 1 **per cent**. The percentage error is therefore 1.

LATIN *per*, for each; *centum*, a hundred.

PERFECT NUMBER

A **number** which is **equal** to the **sum** of its **factors**, including 1 but excluding the **number** itself.
Example: The factors of 6 are 1, 2, 3 and 6. Excluding 6 the sum of the factors is 6 so 6 is a perfect number $(1+2+3 = 6)$. 28 is also a perfect number. The next one is 496.

PERFECT SQUARE

A **number** that can be expressed as the **product** of two **equal whole numbers**. Its **square root** is therefore a whole number.

Examples: $25 = 5 \times 5 \quad 64 = 8 \times 8 \quad 625 = 25 \times 25$.

Thus 25, 64 and 625 are perfect squares. Such numbers can be represented by **squares** or by **dots**.

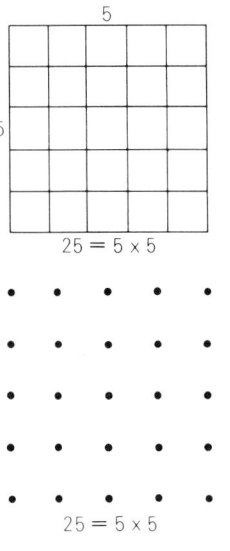

$25 = 5 \times 5$

$25 = 5 \times 5$

PERIMETER

The **length** of the **boundary** of a closed **figure** drawn in a **plane**.

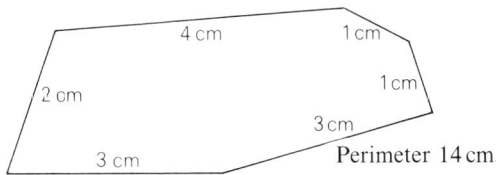

Perimeter 14 cm.

GREEK *peri*, around; *metron*, **measure**.

PERIODIC DECIMAL

See REPEATING DECIMAL.

PERMUTATION

The ways in which things can be arranged or selected. The letters A, B and C can be arranged or permuted in 6 ways, ABC, ACB, BAC, BCA, CAB and CBA. The **order** is important. If A, B and C run a race there are 6 possible arrangements for the 1st and 2nd places. They are AB, AC, BC, BA, CA and CB. If the order does not matter there are only three ways of choosing two children from three (Note AB is then the same as BA).

We could choose AB, AC or BC. When order does not matter the arrangement is called a **combination**.

PERPENDICULAR

A **line** or **plane** that is at **right angles** to another line or plane.

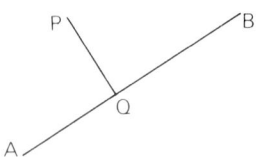

Perpendicular is also used in the sense 'PQ is perpendicular to AB'. Originally the **term** was only used with reference to **vertical** and **horizontal** line.

PERPENDICULAR BISECTOR

A **line** at **right angles** to another line and passing through its **midpoint**.

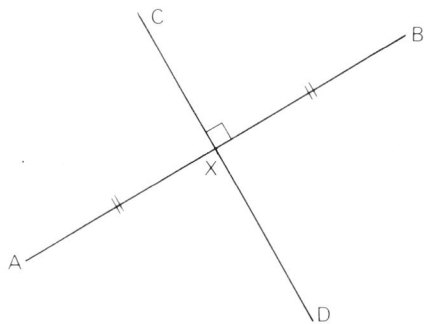

Example: X is the midpoint of AB, **angle** CXB is 90° and CD is the perpendicular bisector of AB.

PERSONAL UNITS

Units that depend on the person using them, such as the **pace**, **span** or **cubit**. Personal units are one form of **arbitrary units**.

PI

The Greek letter π used to **represent** the **ratio** of the **circumference** of a **circle** to its **diameter**; $\dfrac{c}{d} = \pi$.

In Biblical days it was taken as 3 but $3\frac{1}{7}$ and 3.14 are better **approximations**.

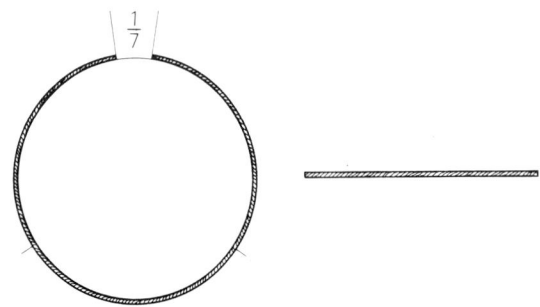

A piece of string the same **length** as the diameter can be fitted three **times** round a circle and there is then a small gap left. This gap is about $\frac{1}{7}$ of the diameter. More **accurate values** can be found from **series** such as $\pi = 4(1 - \frac{1}{3} + \frac{1}{5} - \frac{1}{7} + \frac{1}{9} - \frac{1}{11} + \frac{1}{13} \ldots)$ and electronic **computers** have calculated the value to more than $100\,000$ **decimal** places. The **digits** do not at any time form a recurring **pattern**. The **symbol** π was first used at the end of the 17th **century**. **Archimedes** found pi was between $3\frac{10}{71}$ and $3\frac{1}{7}$, that is between 3.1408 and 3.1429 (approximately). In 1882 it was proved that π could never be written exactly no matter how many decimal places we use. Correct to 20 decimal places pi is 3.141 592 653 589 793 238 46.

GREEK *pei* or *pi*, the sixteenth letter of the Greek alphabet.

PIAGET

A Swiss psychologist who has contributed many books and articles on how children develop, with special reference to how they think. His ideas have led to many changes in the way **mathematics** is taught.

PICTOGRAM

Also called a **pictograph** or picture **graph**. Pictures are used to **represent** objects instead of **points**, **lines** or blocks.

Examples: 1.

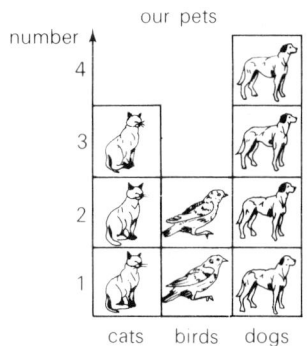

2 For large **numbers** one picture may represent many objects.

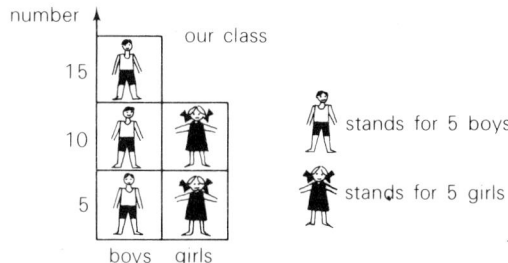

LATIN *pictus*, painted.
GREEK *gramma*, letter or **figure**.

PICTOGRAPH   See PICTOGRAM.

PICTORIAL REPRESENTATION

Any form of **graph**, drawing, or means of showing information by means of a picture or **diagram**. Pictorial representation includes **bar charts**, **block graphs**, **histograms**, line graphs, **pie-graphs**, and **Venn diagrams**.

PIE-GRAPH

Also called **circle graph** or pie-chart. A form of **representation** in which the **frequency** or **amount** of each **quantity** is **proportional** to the **angle** at the **centre** of a **circle**.

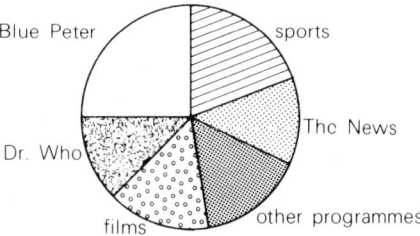

Blue Peter (90°) has twice as many supporters as Dr. Who (45°).

So called due to its resemblance to a meat or fruit pie.
GREEK *graphein*, to write.

PINT

An **imperial unit** for measuring liquids. In the **metric system** liquid would be measured in **litres**.

1 pint is approximately 0.568 litres.

1 litre is approximately 1.76 pints.

LATIN *pingo*, to paint. A mark was painted on the **side** of a container to show where a pint came to.

PLACE HOLDER

1 A **variable** in a mathematical **sentence**.

Examples:

> $\square + 3 = 8$, $2x + 9 = 2$. $\square$ and x are place holders. They hold the place for a **number**.

2 **Zero**, when used with other **digits**, enables us to decide the **value** of these other digits. For instance the 0 in 50 holds the **units** place and enables us to see that the 5 is the number of tens. In 500 the 5 is the number of hundreds and two zeros hold the places of the tens and units.

PLACE VALUE

The **value** of a **digit** due to the **position** or place it occupies.

Example: In 34838 the 3 on the left **represents**

Number of 10 000's.   Number of 10's.

three 10 000's but in the other position the three represents three 10's.

The first 3 has a place value of 30 000 and the second a place value of 30.

PLAN

A **diagram** representing **solid** objects by means of a **plane** drawing, that is one that has only **two dimensions**. In particular it is used for a view as seen from above.

Example:

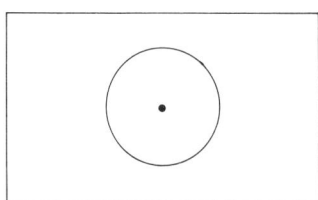

Plan of a **cone** placed on a book.

Elevation is used to describe the objects as seen from the front.

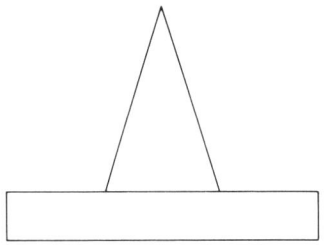

Elevation of cone on a book. Also used loosely for any large **scale map**.

PLANE

A **flat surface**. A plane extends infinitely in all **directions**. More accurately a surface such that if any two **points** on it are joined the **line** so formed lies wholly on the surface.

PLANE FIGURE

Any **figure** which lies in a **plane**.

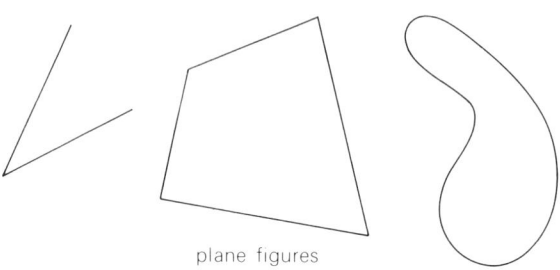

plane figures

Plane figures.

PLANE GEOMETRY

The **geometry** of **figures** that can be drawn in a **plane**.

Example: The study of **circles**, **quadrilaterals** and **triangles**.

Plane geometry deals with **shapes** with **two dimensions** and not those with **three dimensions** such as **cuboids** or **cones**.

PLANE SECTION   See CROSS SECTION.

PLANE TABLE

A table used in **surveying**.

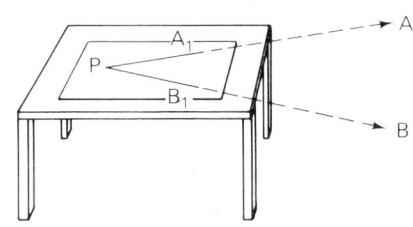

In plane table surveying a sheet of paper is attached to a table and a **point** (P) is marked to **represent** a fixed point from which all **measurements** and **bearings** will be taken. In the diagram a **line** from P towards an object A is drawn and similarly one towards B. PA and PB are measured and PA_1 and PB_1 drawn to **scale** to represent these distances.

Other objects are treated in a similar way to A and B and a survey of the **area** is thus built up.

PLANIMETER   See PANTOGRAPH.

PLATO, 427–347 BC
A great philosopher and teacher who studied the **geometry** of **Pythagoras** and also that of the **Egyptians**. Above the entrance to his school of philosophy was written 'Let no one ignorant of geometry enter my doors'. He laid down the basic principles of **mathematics** insisting upon clear **definitions** and logical **proof**. Plato was especially interested in the mystical properties of **numbers**.
(See PLATONIC SOLIDS.)
The dates of Plato's birth and death are not known for certain.

PLATONIC SOLIDS
See **regular solids**.
Named after the Greek philosopher **Plato**.
LATIN *solidus*, solid.

PLOT, PLOTTING
The process of marking **points** usually given in the form of **coordinates** e.g. (3, 2). These are plotted with reference to two **lines** at **right angles**, called **axes**.

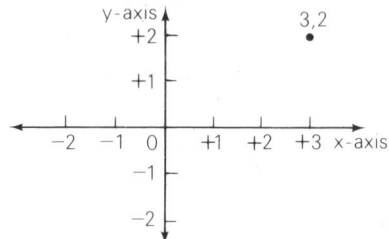

The points may then be joined to form a line or **curve**.
(See GRAPH.)

PLUMB LINE
A thread or string with a **weight** (or plumb-bob) attached.

The **line** of the string is **vertical** and for all practical purposes it is at **right angles** to the **horizontal**. It does not point directly towards the Earth's **centre** due to the fact that the Earth is not a true **sphere** and the materials it is made of are not **uniform**.
LATIN *plumbum*, lead (from the lead weight of the plumb line).
OLD ENGLISH *linum*, flax. (See LINE.)

PLUS
Add

The plus **sign**, + is used to indicate the **operation** of **addition**. It is also called the addition sign. The sign + also denotes a **positive integer** such as $^+4$. This is read as positive four.

The sign was first used in the 15th **century** to mark sacks of grain that were overweight. In the same way sacks that were too light were marked with what we now called the **minus** sign, −.

The plus sign probably arose from the latin 'et' for 'and' which was written & or ϙ and hence +.

LATIN *plus*, more.

P.M. or p.m.

Abbreviation for **post meridiem** or **post meridian**. After midday. Any **time** between noon (12 a.m.) and the following midnight.

POINT

1 **Euclid** defined a point as that which has **position** but no **magnitude**. This mark . obviously has **size** but if it is imagined to grow smaller and smaller we get nearer and nearer to the idea of a point. If thought of as only having position a point can be represented by **coordinates** (the point 5, 2) or as where two **lines intersect**.

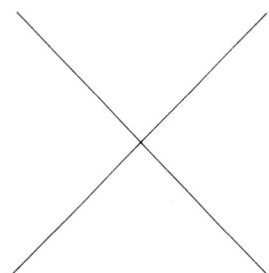

2 Because of the difficulties involved if a point has no size mathematicians leave point as an undefined **term**. That is they use the word and idea but do not attempt to say exactly what it is.

(See DOT.)

POINT SYMMETRY

Balance or **symmetry** about a **point**.
A is joined to O and produced to A' so that AO = OA'

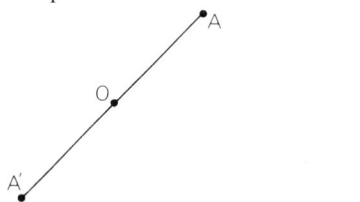

A' is the **image** of A and there is point symmetry about O. This applies to every point on a shape that has point symmetry about a given point O.

Example :

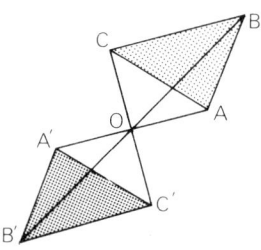

In **triangle** ABC, A **maps** to A', B to B' and C to C'. There is point symmetry about O. Triangle ABC maps on to A'B'C'.

POLYGON

A **plane closed figure** bounded by **straight lines**. Generally thought of as a figure with many **sides** but its name means 'many **angles**'.

Many polygons have special names according to the **number** of sides they have. Here are some examples :

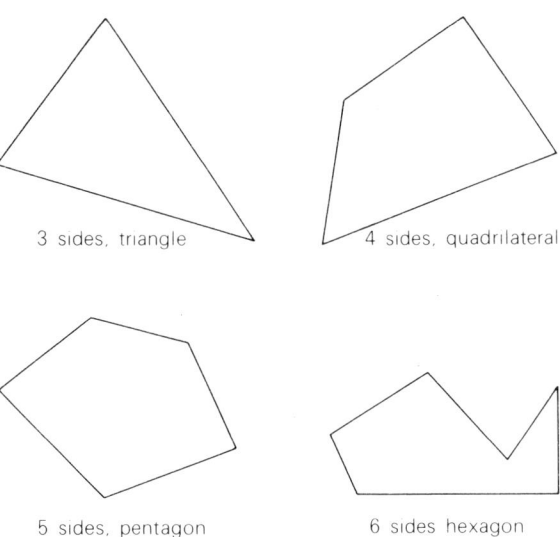

3 sides, triangle

4 sides, quadrilateral

5 sides, pentagon

6 sides hexagon

If the sides of a polygon are all **equal** in **length** and if the angles are all equal in **size** the polygon is **regular**.
If these conditions are *not* satisfied it is **irregular.**
GREEK *poly*, many ; *gonia*, angle.

POLYGONAL NUMBERS

Another name for **figurate numbers**. If **dots** can be arranged to make **polygons** according to a certain rule given below then the **number** of dots is called a polygonal number. **Triangular** and **square numbers** are particular cases since **triangles** and **squares** are polygons. Consider pentagonal numbers (penta, five).

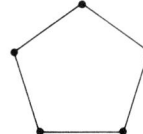

A **pentagon** with **sides** twice the **length** is now built round the first pentagon.

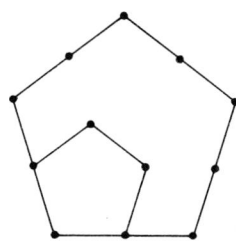

It has 10 dots (.) on its **perimeter**. The **total** number of dots is 12. 12 is therefore the next pentagonal number. Building round the outside of the last figure it would be found that 22 is the next pentagonal number. These numbers can be found more rapidly from a number **pattern**.
If we start with 1 we have:

1, 1 + 4, 1 + 4 + 7, 1 + 4 + 7 + 10 ...

1,   5,   12,   22

The next will be $1 + 4 + 7 + 10 + 13$, that is 35. Adding 3 to the last number 13 we get 16 which is the last **term** of the next **sequence**. $1 + 4 + 7 + 10 + 13 + 16 = 51$ so that is the next polygonal number.
GREEK *poly*, many; *gonia*, **angle**.

POLYHEDRON
A **solid** bounded by **polygons**.

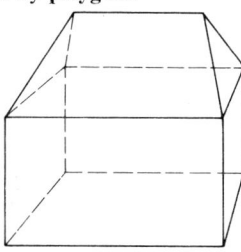

GREEK *poly*, many; *hedra*, seat (**face**).

POLYOMINOES
Figures formed by **equal squares** that meet along a complete **edge**.

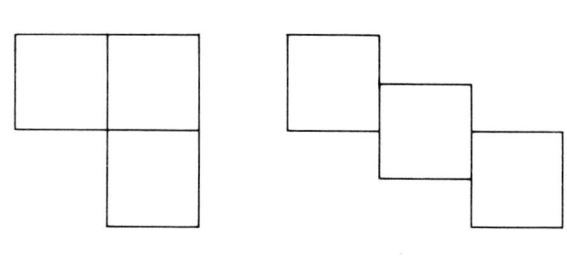

This is a polyomino.   This is not a polyomino.

A polyomino formed by two squares is called a **domino**.

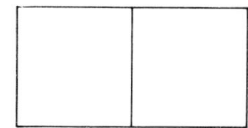

Trominoes are formed from three squares. There are two trominoes.

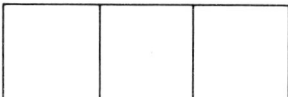

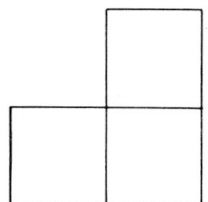

All others can be formed from these by turning. Also see TETROMINOES and PENTOMINOES.
GREEK *poly*, many.

POPULATION
The **term** used in **statistics** to denote the whole **set** from which a **sample** is taken.
Example: A factory might test one in every hundred light bulbs that it makes. The population refers to all the bulbs from which those sampled are selected.

POSITION
The place occupied.

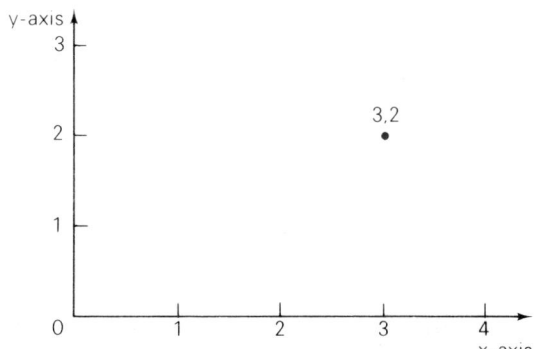

The position of a **point** can be indicated by its **coordinates**.

Another method is by means of a position **vector**.
Example:

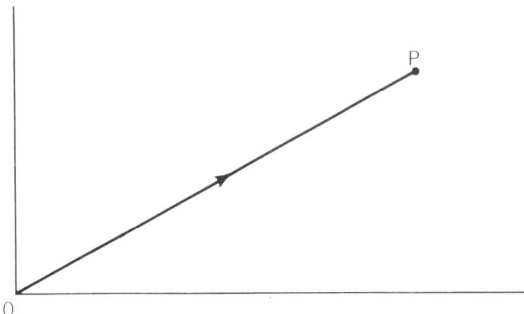

The **direction** and **length** of the **vector** determine the position of the point P.

POSITIVE NUMBERS
Numbers with a **plus sign** (+) written before them, such as +4 or preferably $^+4$. (This is read as positive four). Positive and **negative numbers** can be represented by **points** on a **line**, the **direction** being indicated by the + or − sign.

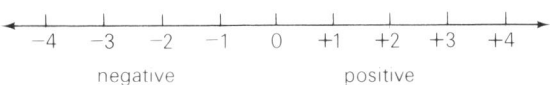

POST MERIDIEM or POST MERIDIAN
See P.M. or p.m.
LATIN *post*, after; *meridies*, midday.

POUND
1 A **unit** of money. £1 = 100p.

2 A unit of **mass** (1 lb mass) and of **weight** (1 lb weight). With metrication these are replaced by the **kilogram**. 1 kilogram is approximately 2.2 pounds.

POWER
1 We read 5^4 as the 4th power of 5, or 5 to the power of 4 5^4 is an **abbreviation** for $5 \times 5 \times 5 \times 5$. The power shows how many **equal numbers** have been multiplied together, $a \times a \times a = a^3$, that is a to the power of 3.

2 The **rate** at which **work** is done. The power of an engine is calculated from the amount of work it can do in a given **time**.

PRACTICE
1 **Simple practice.**
A method of calculating the cost of a **number** of articles when the price of one is known.

Example: Find the cost of 362 articles at £1·67 each.

362 at £1	cost £362		
362 at 50p	cost £181	$(50p = \frac{1}{2}$ of £1)	
362 at 10p	cost £ 36·20	$(10p = \frac{1}{10}$ of £1)	
362 at 5p	cost £ 18·10	(5p $= \frac{1}{2}$ of 10p)	
362 at 2p	cost £ 7·24	(2p $= \frac{1}{5}$ of 10p)	
Total	£604·54		

2 Compound practice.
A method of finding the cost of a **quantity** when the price of a **unit** of that quantity is known.
Example: Find the cost of 2 kg 240 g at £1·75 for 1 kg.

Cost of 1 kg	= £1·75	
Cost of 1 kg	= £1·75	
Cost of 200 g =	35p	$(200 g = \frac{1}{5} \times 1$ kg
Cost of 40 g	7p	$(40 g = \frac{1}{5}$ of 200 g)
Total	= £3·92	

PRESSURE
The **force** exerted. Pressure is measured by the force on each **unit area**. As a diver goes deeper in the water so the pressure of the water above him increases.

PRIME FACTOR
A **prime number** that is a **factor** of a given **number**.
Example: 2 and 3 are prime factors of 36. 6 and 12 are factors of 36 but as they are not prime numbers they cannot be prime factors.
LATIN *prima*, first; *facere*, to make.

PRIME NUMBERS
A **natural number** that has only two **factors**, 1 and itself. 1 is not regarded as a prime number as it has only one factor, that is 1.
Examples: 2, 5, 11, 37, ...

PRINCIPAL
1 The money borrowed or invested. **Interest** is paid on this and the principal **plus** interest is called the **amount**.
Example: A man invests £500 at 8 **per cent**. What is the amount after one **year**?
Principal = £500.
$$\text{Interest} = \frac{£8 \times 500}{100} = £40$$
Amount = Principal + Interest = £500 + 40 = £540.
See SIMPLE INTEREST and COMPOUND INTEREST.

2 The one of greatest importance.
Example:

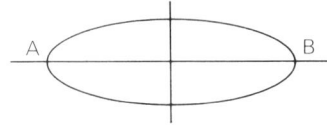

A B is the principal or **major axis** of the **ellipse**.

PRISM

A **solid** with two **parallel faces** that are **polygons**. The other faces are **parallelograms**.
Examples:

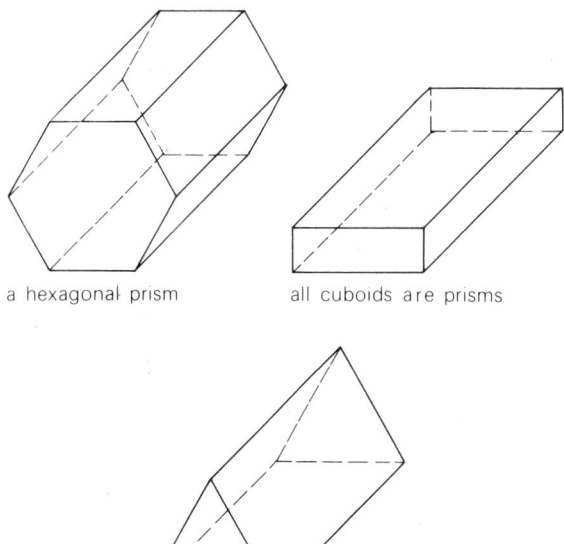

a hexagonal prism   all cuboids are prisms

a triangular prism

How much do 38 nails **weigh** if they are each **4 grams**? This is a problem for some people as they must first decide whether to **add**, **subtract**, **multiply** or **divide**. For a person who knows right away that they must multiply it is not a problem. It follows that what is a problem for one person is not a problem for another.

PRODUCE

To extend or make longer.

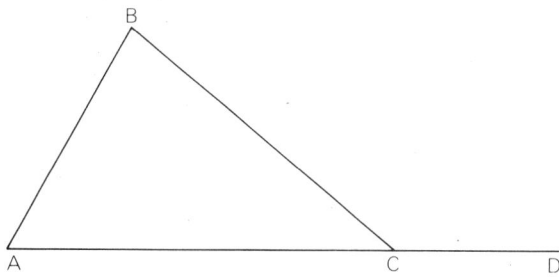

The **side** AC of the **triangle** ABC has been produced to D.

PRODUCT

The **result** of multiplying two or more **numbers**.
See also **Cartesian product**.

PROFIT

The **amount** for which an article is sold **minus** the amount originally paid for it.
The profit is therefore **selling price − cost price** or SP − CP.
Example: A man bought some potatoes for £2 and sold them for £3.
Profit = SP–CP = £3 − £2 = £1.
(See also COST PRICE.)

PROBABILITY

The likelihood of an event happening. The probability of a 6 when a **die (dice)** is thrown is one chance out of 6. This probability is written as $\frac{1}{6}$.

$$\text{Probability} = \frac{\textbf{Number of favourable results.}}{\textbf{Total number of possible results.}}$$

Example:
Find the probability of getting **less than** 3 when throwing a dice.

Probability $= \frac{2}{6} = \frac{1}{3}$ (1 and 2 are favourable. There are six possible results.)

If an event cannot happen the probability is 0. (The probability I will **weigh** 2 **tons**.) If an event is certain to happen the probability is 1. (The probability that you have a mouth.)

PROBLEM

A task requiring reasoning and not capable of being done by a remembered technique alone. 38 × 4 is not a problem.

PROJECT

1 A task that requires combining skills. Projects normally take a considerable **amount** of **time**, as for instance building a model fort or doing a survey of local shopping habits.

2 To transform **points** from one **shape** or picture to another. For instance to enlarge.

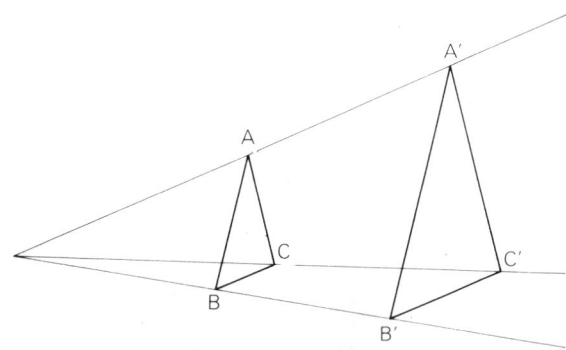

Triangle ABC has been enlarged to form triangle A'B'C'.

PROJECTILE

Something which is thrown, hurled, fired or projected.
Examples: A stone, bullet or rocket.
LATIN *projectum*, to throw forward.

PROOF

A logical series of **statements** that establish the truth of a proposition.

PROPER FRACTION

A **fraction** in which the numerator is **less than** the **denominator**.

PROPER SUBSET

A proper subset of **set** A is any **subset** of A that has fewer **elements** than set A.
Example: Set A = {dog, cat, canary}. {dog, cat}, {dog} are proper subsets but {dog, cat, canary} is **not** a proper subset.
{dog, cat, canary} is called a subset of set A.

PROPERTY

Any **attribute**, such as colour, **size**, **number** of **sides** or any other characteristic.
Example: Properties (or attributes) of a pencil may include: cylindrical **shape** before sharpening, **cone** shaped end where sharpened, made of wood and lead, suitability for writing with, colour.

PROPORTION

See DIRECT PROPORTION and INVERSE PROPORTION.

PROTRACTOR

An **instrument** for measuring **angles**.

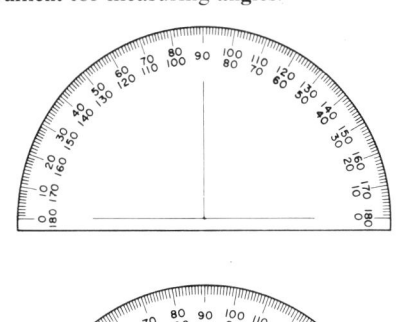

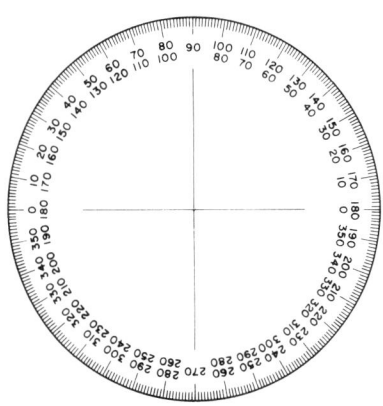

PULLEY

A device for raising **masses** or changing the **direction** in which **power** is applied. In its simplest form a pulley consists of a wheel that turns about an **axis**, with a rope or chain in a groove on the rim of the wheel.

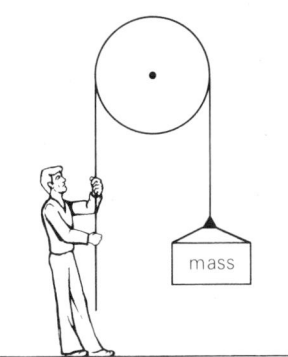

The man provides a downward **force** and this is changed to an upward force on the mass. There are many systems of pulleys so arranged that a small force, moving a large distance can move a large mass a small distance.

PUNCHED CARDS

Cards on which information can be stored by cutting from the edge of the card to the punched holes so as to indicate one of two choices. The cards can then be used to sort information, some being designed for feeding into a **computer.**

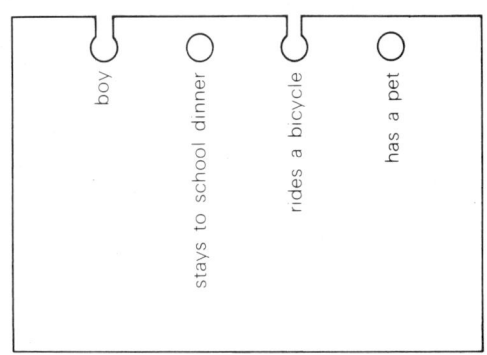

Example: On the card shown David has indicated he is a boy, rides a bicycle but does not stay to school dinners and does not have a pet.

The opposite method could be used. The part that is not cut then shows the **property** chosen. David would therefore cut out the 2nd and 4th instead of the 1st and 3rd sections.

PURE MATHEMATICS

Mathematics that generally speaking does not have a practical application. It is studied for its own sake rather than to **solve problems** in science or an industry.
Mathematics which is not **applied mathematics**.

PYRAMID

A **solid** with a **polygon** as its **base**, the other **faces** being **triangles** meeting at a **vertex**.

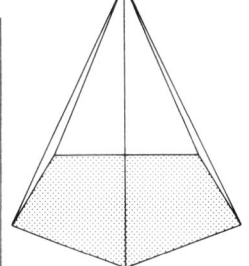

A pyramid with a pentagonal base.

An Egyptian pyramid.

The brotherhood was then banned by the government. Pythagoras found **relations** between **mathematics** and music but is now best known for the **theorem** that bears his name. The **Theorem** of Pythagoras states that for a **right-angled triangle** the **sum** of the **squares** on the two **sides** containing the **right angle** is **equal** to the square on the side opposite the right angle, (the **hypotenuse**).

PYTHAGORAS

Pythagoras was born on the Greek Island of Samos sometime around 570 BC and died when he was about 65 **years** old.

He studied in Egypt and later in Babylon finally settling in Cretona, a Greek colony in southern Italy. Here he formed a secret society called the Pythagorean brotherhood. **Members** recognised each other by secret **signs** and badges including a pentagonal star

They took an oath not to marry each other (but Pythagoras married one of the society), not to wear wool, never to touch a white cockerel and not to poke a fire with an iron poker. They dabbled in magic and religion. Some were put to death for giving away the society's secrets. The Pythagoreans became powerful but were hated by the people of Cretona who were not members of the brotherhood. This led to the school of Pythagoras being burnt to the ground and many of the society murdered.

Q

QUADRANGLE

The word means four **angles**.

More accurately it is a **plane**, **closed figure** formed by joining four **points**, no three of these being in a **straight line**.

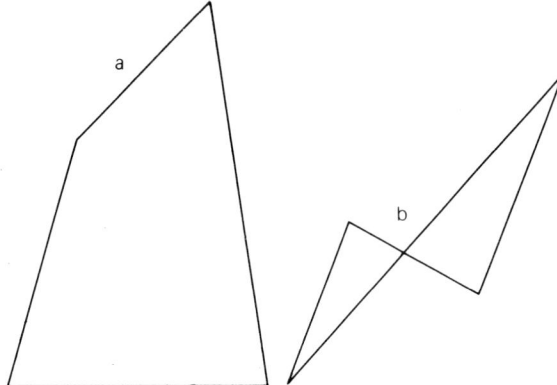

a is also called a **quadrilateral**. Some mathematicians also call *b* a quadrilateral but the **term** is best used only for cases where the four **lines** do not cross.

A complete quadrangle consists of all six lines formed by joining the four points in **pairs**.

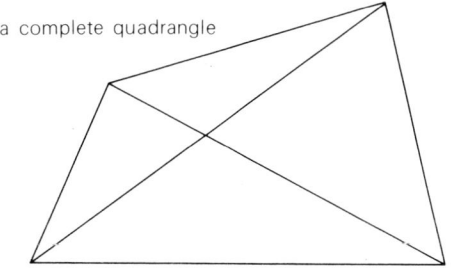

a complete quadrangle

LATIN *quadri*, four; *angulus*, an angle.

QUADRANT

1 Any one of the four parts into which a **plane** is divided by two **axes** at **right angles** to each other. The quadrants are named as shown in the **diagram.**

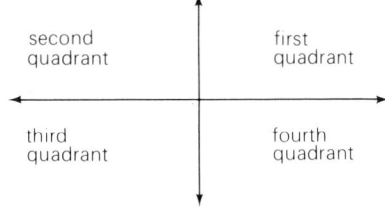

2 A quadrant of a **circle** is that part bounded by two **radii** at right angles to each other and the **corresponding arc** joined their end **points**.

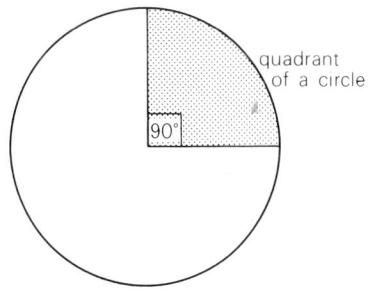

quadrant of a circle

90°

3 An instrument for measuring the **sun's angle of elevation**. It can also be used for other angular **measurements**.

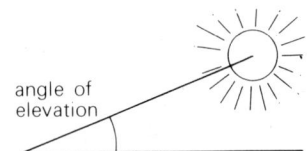

angle of elevation

LATIN *quadrans*, a fourth part, a **quarter**.

QUADRILATERAL

The word means four **sides**. More accurately a **plane, closed figure** with four sides, or a four sided **polygon**. **Kite, rectangle, rhombus, parallelogram, square,** and **trapezium** are all special cases of **quadrilaterals**.

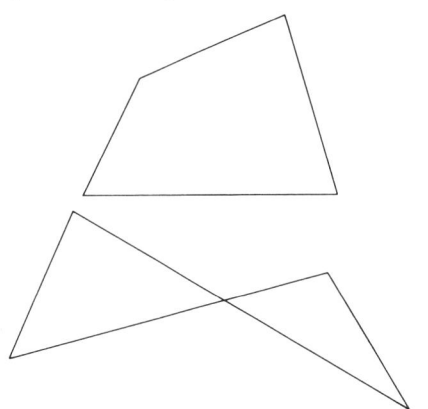

Some mathematicians would call this a quadrilateral.

The **term** is best used only for cases where the four **lines** do not cross. A complete quadrilateral consists of all six lines formed by joining four **points**, no three of which are in a **straight line**.

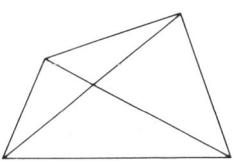

(See QUADRANGLE.)
LATIN *quadri*, four; *lateris*, side.

QUANTITY

An **amount**.
Example: The quantity of water in a jug.

QUART

1 An **imperial unit** of **capacity** (**liquid measure**).

1 quart = 2 **pints**.   1 quart = $\frac{1}{4}$ **gallon**.

1 quart is approximately 1.14 **litres**.

2 An imperial unit of capacity (**dry measure**).
1 quart = 2 pints.   8 quarts = 1 **peck**.
LATIN *quartus*, a fourth, (a **quarter** of a gallon).

QUARTER

1 One of four **equal** parts.

2 In **imperial units** 28 lb is called a quarter. It is a quarter of a **hundredweight** (1 cwt = 112 lb).
3 A **fraction** with 1 as **numerator** and 4 as **denominator**.
LATIN *quartus*, a fourth.

QUINTAL

A **unit** of **weight** or **mass** in the **metric system**.
1 quintal = 100 **kilograms**.
Formerly used for a **hundredweight**, that is 112 **pounds**.

QUOTIENT

The **number** resulting from a **division**.
Example: $8 \div 2 = 4$. 4 is the quotient. Quotient is also used for an expression such as $8 \div 2$ or $\frac{8}{2}$.

When a **remainder** occurs the **whole number** resulting from the division may be called the quotient.
Example: $9 \div 2 = 4$ rem 1. 4 is called the quotient. Note: There are several interpretations.
In the last example 4, $9 \div 2$, $\frac{9}{2}$ or $4\frac{1}{2}$ may be referred to as the quotient.

QUOTITION

One aspect of **division**, the other being **partition**. In quotition or **grouping** we find how many **groups** of a stated **size** are contained in a given **number**.
Example: How many groups of 2 are contained in 8?
$8 \div 2 = 4$.

R

RADIUS, RADII
The radius of a **circle** is a **line** from the **centre** to a **point** on the **circumference**.

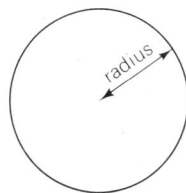

Radii is the plural of radius. The radius of a **sphere** is a line from the centre of the sphere to any point on its **surface**. LATIN *radius*, a **rod** or spoke.

RANDOM SAMPLE
A **term** in **statistics** denoting part of a **population** that is chosen to **represent** the whole, every **member** having an **equal** chance of being selected.

RANGE
1 In **statistics** the **difference** between the greatest and **least** **values** in any one **set** of **data**.
Example: The oldest and youngest people are 85 and 19 years old. The range is 66 years.

2
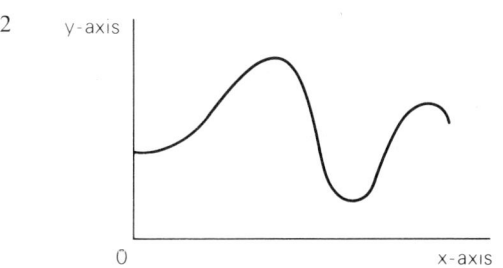

In any **function** there are two sets of values, denoted on the **graph** by x and y. The values of y form the range (also called co-domain). (Those of x form the **domain**. When in the form of **ordered pairs**, such as $(3, 2)$, the set of values of the second **number** of the **pair** form the range.)

3 The values of a **variable** in any mathematical **statement**. This interpretation is not so accurate as 2.

RATE
1 The **amount** of change of one **quantity** with respect to another.

2 The **interest**, generally stated as a **percentage** for a **year**. *Example:* A rate of 5 **per cent** per year.

3 Rates. Money paid by shopkeepers, householders and others to pay for local costs such as education, water and road maintenance.

RATIO
A **comparison** of two **quantities** by expressing one as a **fraction** of the other.
Example: The ratio of 2 **mm** to 1 **cm** can be expressed as $2\,\text{mm} : 1\,\text{cm}$, $2\,\text{mm}/1\,\text{cm}$ or $\dfrac{2\,\text{mm}}{1\,\text{cm}}$ and these can be simplified to $1:5$, $1/5$ or $\frac{1}{5}$ respectively.

RATIONAL NUMBER
A **number** that can be expressed as a **fraction** or **ratio**, that is as a fraction in which the **numerator** and **denominator** are **whole numbers** but the denominator is not **zero**.
Example: $\frac{3}{4}$, 0.4 since $\frac{4}{10} = \frac{2}{5}$.
An **irrational number** is one that is not rational.

RAY
A **half-line**.

A **line** extends indefinitely and therefore consists of two rays.

READY RECKONER

A **table**, or book of tables, in which the results of a **calculation** can be read off directly.

No.	£ p 0 20	£ p 0 21	£ p 0 22	No.	£ p 0 20	£ p 0 21	£ p 0·02
1	0 20	0 21	0 22	51	10 20	10 71	11 22
2	0 40	0 42	0 44	52	10 40	10 92	11 44
3	0 60	0 63	0 66	53	10 60	11 13	11 66
4	0 80	0 84	0 88	54	10 80	11 34	11 88
5	1 00	1 05	1 10	55	11 00	11 55	12 10
6	1 20	1 26	1 32	56	11 20	11 76	12 32
7	1 40	1 47	1 54	57	11 40	11 97	12 54
8	1 60	1 68	1 76	58	11 60	12 18	12 76
9	1 80	1 89	1 98	59	11 80	12 39	12 98
10	2 00	2 10	2 20	60	12 00	12 60	13 20
11	2 20	2 31	2 42	61	12 20	12 81	13 42
12	2 40	2 52	2 64	62	12 40	13 02	13 64
13	2 60	2 73	2 86	63	12 60	13 23	13 86
14	2 80	2 94	3 08	64	12 80	13 44	14 08
15	3 00	3 15	3 30	65	13 00	13 65	14 30
16	3 20	3 36	3 52	66	13 20	13 86	14 52
17	3 40	3 57	3 74	67	13 40	14 07	14 74
18	3 60	3 78	3 96	68	13 60	14 28	14 96
19	3 80	3 99	4 18	69	13 80	14 49	15 18
20	4 00	4 20	4 40	70	14 00	14 70	15 40
21	4 20	4 41	4 62	71	14 20	14 91	15 62
22	4 40	4 62	4 84	72	14 40	15 12	15 84
23	4 60	4 83	5 06	73	14 60	15 33	16 06
24	4 80	5 04	5 28	74	14 80	15 54	16 28
25	5 00	5 25	5 50	75	15 00	15 75	16 50
26	5 20	5 46	5 72	76	15 20	15 96	16 72
27	5 40	5 67	5 94	77	15 40	16 17	16 94
28	5 60	5 88	6 16	78	15 60	16 38	17 16
29	5 80	6 09	6 38	79	15 80	16 59	17 38
30	6 00	6 30	6 60	80	16 00	16 80	17 60
31	6 20	6 51	6 82	81	16 20	17 01	17 82
32	6 40	6 72	7 04	82	16 40	17 22	18 04
33	6 60	6 93	7 26	83	16 60	17 43	18 26
34	6 80	7 14	7 48	84	16 80	17 64	18 48
35	7 00	7 35	7 70	85	17 00	17 85	18 70
36	7 20	7 56	7 92	86	17 20	18 06	18 92
37	7 40	7 77	8 14	87	17 40	18 27	19 14
38	7 60	7 98	8 36	88	17 60	18 48	19 36
39	7 80	8 19	8 58	89	17 80	18 69	19 58
40	8 00	8 40	8 80	90	18 00	18 90	19 80
41	8 20	8 61	9 02	91	18 20	19 11	20 02
42	8 40	8 82	9 24	92	18 40	19 32	20 24
43	8 60	9 03	9 46	93	18 60	19 53	20 46
44	8 80	9 24	9 68	94	18 80	19 74	20 68
45	9 00	9 45	9 90	95	19 00	19 95	20 90
46	9 20	9 66	10 12	96	19 20	20 16	21 12
47	9 40	9 87	10 34	97	19 40	20 37	21 34
48	9 60	10 08	10 56	98	19 60	20 58	21 56
49	9 80	10 29	10 78	99	19 80	20 79	21 78
50	10 00	10 50	11 00	100	20 00	21 00	22 00
¼	0·05	0·0525	0·055	¾	0·15	0·1575	0·165
½..	0·10	0·1050	0·110	1¼..	0·25	0·2625	0·275

REAL NUMBER

Any **rational number** or **irrational number**.
The **set** of real numbers is denoted by R.
Examples: 6, 2.9, $\frac{3}{4}$, $\sqrt{2}$.

RECIPROCAL

A **number** which, multiplied by a given number **results** in 1.
Examples: $2 \times \frac{1}{2} = 1$ so $\frac{1}{2}$ is the reciprocal of 2. Also 2 is the reciprocal of $\frac{1}{2}$. $\frac{3}{5}$ is the reciprocal of $\frac{5}{3}$ since $\frac{3}{5} \times \frac{5}{3} = 1$. Also $\frac{5}{3}$ is the reciprocal of $\frac{3}{5}$.
The reciprocal is also called the multiplicative **inverse**.

RECORD, RECORDING

Writing, drawing or other **representation** of information.

RECTANGLE

A **quadrilateral** with four **right angles**. (From this information it can be proved that the opposite **sides** are **equal**).

rectangle

A rectangle is frequently referred to as an **oblong**. They are not however the same – a **square** is a rectangle, but it is *not* an oblong.
LATIN *recti*, upright; *angulus*, angle.

RECTANGULAR NUMBERS

Numbers that can be represented by **dots** arranged to form a **rectangle**.
Examples:

6

15

Square numbers are special cases of rectangular numbers so that 9, 16, 25, 36, ... are square and rectangular numbers. (Note: A **square** is a special form of rectangle.)
LATIN *recti*, upright; *angulus*, **angle**.

RECTILINEAR FIGURE

Any **figure** that is bounded by **straight lines**. A **polygon**.
LATIN *recti*, upright; *linum*, flax (see LINE).

RECURRING DECIMAL

See REPEATING DECIMAL.

REDUCTION

1 The expressing of a **fraction** in its simplest form. See LOWEST TERMS.
Example: $\frac{6}{15}$ can be reduced to $\frac{2}{5}$.

2 Expressing one **quantity** in **terms** of another. The term 'reduce' is not recommended because the fraction is still the same size. It is not reduced (made smaller). It would be better to say, '$\frac{6}{15}$ can be simplified to $\frac{2}{5}$' or '$\frac{6}{15}$ expressed in its lowest terms is $\frac{2}{5}$'.
Example: 2 m 3 cm can be reduced to 203 cm.

3 **Amount** by which a quantity is made smaller.
Example: A shopkeeper may reduce the price of a book from 98p to 90p. The reduction is 8p.

RE-ENTRANT ANGLE

An **interior angle** of a **figure** that is **greater than** 180°. See REFLEX ANGLE.

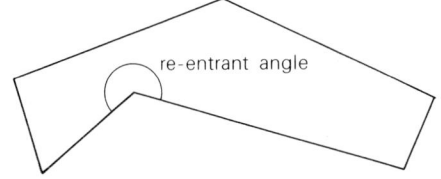

re-entrant angle

REFLECTION

A **mirror image**. See BILATERAL SYMMETRY.

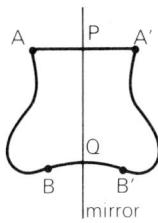

A has a mirror image A', B has an image B'. For any **point** on the **curve** joining A to B there is an image that is **equidistant** from the mirror. For example AP = PA', BQ = QB'. Reflection is a geometric **transformation**.

REFLEX ANGLE

An **angle** that is **greater than** 180° but **less than** 360°.

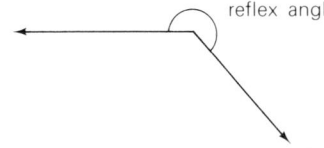

reflex angle

REFLEXIVE PROPERTY OF RELATIONS

A **relation** in which an **element** is related to itself.
Example: John is the same age as John. This is true so the **relation** 'is the same age as' is reflexive. 'is as tall as' [Mary is as tall as Mary]. 'likes the same books as' [Alan likes the same books as Alan]. 'is taller than' is *not* reflexive since it is false to say that Bill is taller than Bill.

REGION

1 For a **plane**. The part of a plane that is inside a simple **closed curve**.

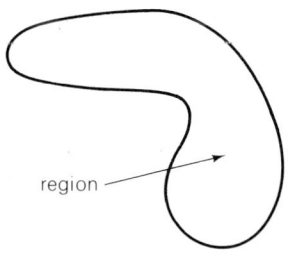

region

2 **For a solid**. The part of the solid that is bounded by a closed **surface**.
Example: The inside of a **sphere** or **cube**.

REGROUP

1 Changing from one **set** of **units** to another. Twenty can be considered as two tens or changed to twenty units (or ones).

Example: $\overset{3\ 10}{\cancel{4}3}$ The 4 tens and 3 units can be re-
 -19 grouped into 3 tens and 13 units

This is the **decomposition method** of **subtracting**.

2 To place in **groups** in a different way.
Example: (See ASSOCIATIVE.) 16 + 18 + 2 can be regrouped as 16 + (18 + 2) or as (16 + 18) + 2. Similarly with **multiplication**: 28 × 6 × 5 can be regrouped as 28 × (6 × 5) or as (28 × 6) × 5.

3 Regrouping may be used for any allowable change in **order** of **operation**.
Example: 27 + 78 + 73 + 22 can be regrouped as
(27 + 73) + (78 + 22) = 200.
This form of regrouping makes use of both the **commutative** and the **associative properties** of addition.

REGULAR POLYGON

A **polygon** with all its **sides** the same **length** and all its **angles equal**.

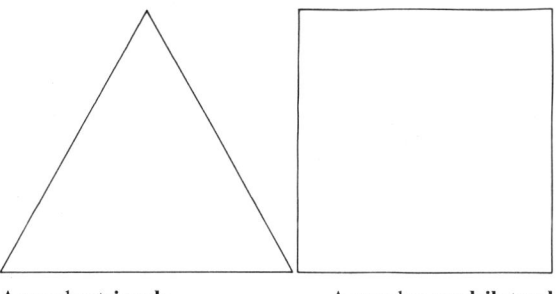

A regular **triangle** (**equilateral** triangle)1   A regular **quadrilateral** (**square**)

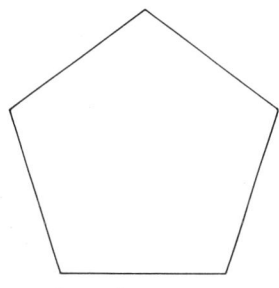

A regular **pentagon**

REGULAR SOLID

Also called **platonic solid** or regular **polyhedron**. A **solid** with **faces** that are **congruent regular polygons**, any two adjoining ones being inclined at the same **angle**. There are five regular solids:

(i) **tetrahedron** (4 **equilateral triangles**);
(ii) **hexahedron** or **cube** (6 **squares**); See HEXA.
(iii) **octahedron** (8 **equilateral triangles**);
(iv) **dodecahedron** (12 **pentagons**) shown below;

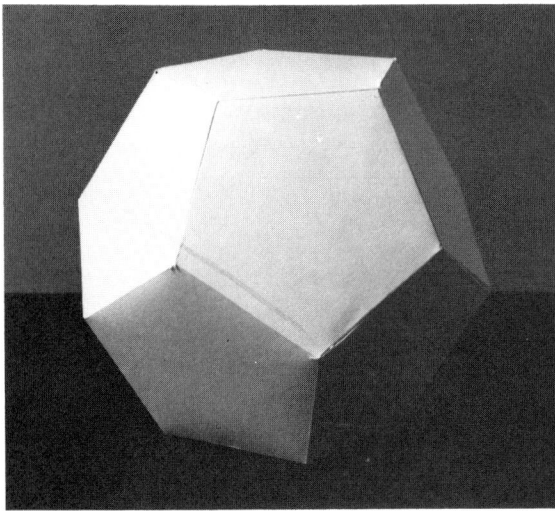

(v) **icosahedron** (20 equilateral triangles).

RELATION

Also called relationship.

1 A connection between **elements** or **members** of a **set** or sets.
Example: 'is **half** of'. 4 is half of 8, 12 is half of 24. 'is the brother of'. John is the brother of Alf.

2 A set of **ordered pairs**.
Example: (4, 8), (12, 24). (8, 16) ... these show the relation 'is half of'.
A relation can be represented by **arrow lines**.
Example:

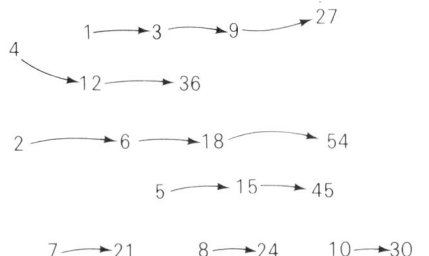

This **diagram** shows the relation 'is one **third** of' for some of the **natural numbers less than** 55.
(See EQUIVALENT RELATION, REFLECTIVE PROPERTY OF RELATIONS, SYMMETRIC PROPERTY OF RELATIONS.)

RELATIVE SPEED

The **speed** of one moving body with respect to another.
Example:

The trains are moving towards each other at speeds of 50 and 20 **metres** per **second**. To a passenger in one train it might seem as if he was not moving and the other train was approaching him at 70 m per second. Thus the relative speed is 70 m per second. If the trains had been travelling in the same **direction** the relative speed would be 30 m per second.

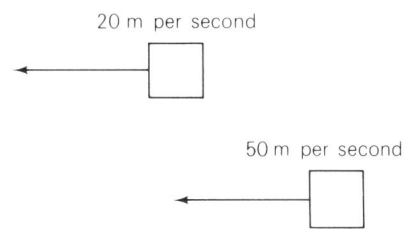

Similarly we have relative **distance**, relative **velocity** and relative **acceleration**.
LATIN *relatum*, to bring back.
OLD ENGLISH *sped*, speed.

REMAINDER

The **amount** left when one **number** is divided by another.
Example:

$$\begin{array}{r} \text{divisor} \quad 25 \to \textbf{quotient} \\ 15\,)\,\overline{387} \to \textbf{dividend} \\ 30 \\ \overline{87} \\ 75 \\ \overline{12} \to \text{Remainder} \end{array}$$

RENAME

A **number** may be expressed in many **equivalent** ways, all of which are different names for that number.

Examples: $\frac{3}{5}$ can be renamed as $\frac{6}{10}$. $4(2+8)$ can be renamed as 40.

REPEATING DECIMAL

Also called **recurring decimal**.

A **decimal fraction** in which one or more **digits** are repeated indefinitely.

Example:

1.3888888 ... To denote that the 8 repeats indefinitely it is written as $1.3\dot{8}$.

$21.18373737... = 21.18\dot{3}\dot{7}$.

$9.215215215... = 9.\dot{2}1\dot{5}$. The · is written over the first and last digits that are repeated.

REPLACEMENT SET

The **set** of **values** that replace a **variable** in an **equation** or in any other mathematical **statement**.

Example: Find the value of $5x+2$ for the replacement set $\{1,2,3\}$.

If $x = 1$. $5(1)+2 = 7$. $x = 2$. $5(2)+2 = 12$. $x = 3$. $5(3)+2 = 17$.

REPRESENT (REPRESENTATION)

Any means of showing a **set** of **values**. **Graphs** and **diagrams** are two important forms of representation.

REPRESENTATIVE FRACTION

A method of stating the **scale** of a map.

Example: A representative fraction of $\frac{1}{10\,000}$ means that any distance on the map represents 10 000 **times** that distance on the ground.

A scale of $\frac{1}{10\,000}$ could also be written as $1:10\,000$ or 1 to 10 000.

RESULT

The last stage in a **proof** or **calculation**.

RESULTANT

A **vector** which is **equivalent** to two or more given vectors.

Example:

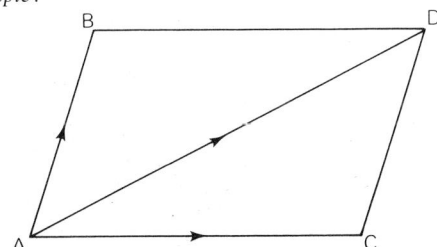

$\overline{AB}$ and $\overline{AC}$ represent two **forces**. Their resultant is represented by the **diagonal** $(\overline{AD})$ of the **parallelogram** ABDC.

REVERSE

Opposite way round.

Examples: Writing the **digits** of 385 in reverse **order** we get 583.

Reverse **operations** are those in which the effect of one is cancelled by the other. Multiplying by 8 is the reverse operation to dividing by 8. Adding 3 is the reverse operation to subtracting 3.

REVOLUTION

A complete turn about a **point** or **axis**.

Example: The **minute hand** of a **clock** makes one revolution every **hour**.

RHOMBUS

A **parallelogram** with two **adjacent sides equal**. From this information it can be proved that all four sides are equal.

RIGHT

1 A **direction**; opposite to left.

2 Erect, as in a right **cone**.

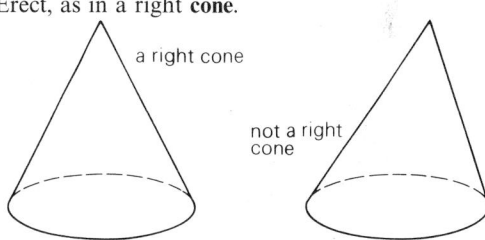

a right cone

not a right cone

RIGHT ANGLE

An **angle** of 90°. One **quarter** of a complete turn or **revolution**.

Examples:

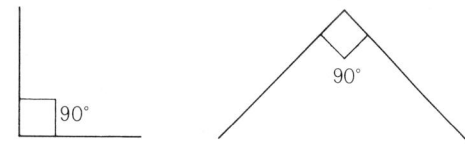

90°

90°

LATIN *rectus*, **straight**, **right**.
OLD ENGLISH *angul*, hook.

RIGHT-ANGLED TRIANGLE

Sometimes called a **right triangle**. A triangle with a **right angle**.

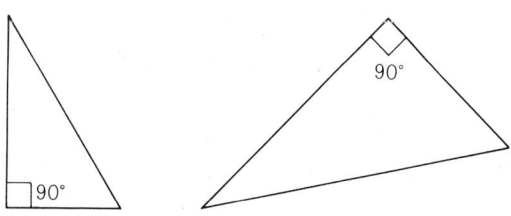

90°

90°

Right-angled triangles.

RIGID

Not able to be distorted. The distance between any two **points** of the body remains **constant**.

Examples: A framework made of **rods** or pipes is rigid if they form **triangles**.

A framework consisting of four rods in the **shape** of a **quadrilateral** is *not* rigid.

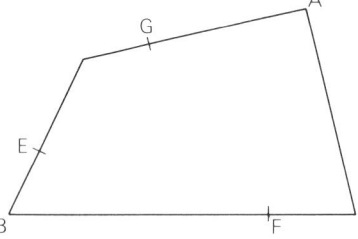

An extra rod joining A to B would make the framework rigid. There are many other **solutions** such as joining E to F. Some of these do not make triangles, for example joining G to F.

RING

1 The **region** of a **plane** between two **concentric circles**.

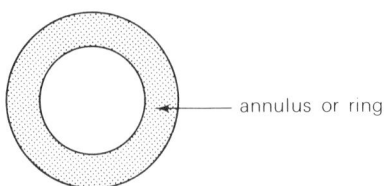

This is more accurately called an **annulus**.

2 The **boundary line** of a circle **or** loop. **Definitions** 1 and 2 are vague and best avoided.

3 A **structure** in advanced **mathematics**.

4 An anchor-ring is another **term** for **torus**.

ROD

1 Pieces of wood or other materials that are used for calculating and examining **relations**. Examples are **Cuisenaire, Colour Factor, Stern**. **Napier's Bones** are also called **Napier's Rods**.

2 A **unit** for measuring **length** in the **imperial system**. 1 rod = $5\frac{1}{2}$ **yards**. A rod is also called a pole or perch.
In the Middle Ages it was fixed by lining up 16 men outside church on a Sunday morning and measuring the combined length of all their left feet.

ROLL

A motion of one **surface** over another without slipping.

ROMAN ABACUS
The illustration shows a simplified form of an **abacus** as some had extra grooves for **calculations** with **fractions**.

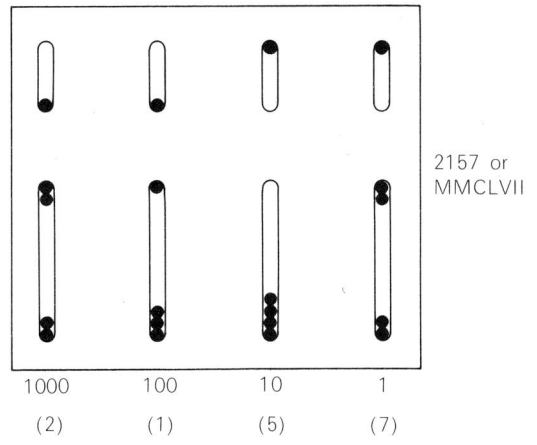

2157 or
MMCLVII

1000 100 10 1
(2) (1) (5) (7)

The abacus consisted of a wax tablet with **pairs** of grooves, one short and one long. 1 bead was placed in the short groove and 4 beads in the long one. Reading from right to left the pairs of grooves were used for 1's, 10's, 100's and 1000's. Beads at the top of a groove represented **numbers**, the one in the short groove standing for 5.

In the example:
the number of 1's is $5+2$,
in **Roman numerals** VII

the number of 10's is 5, that is 50 or L

the number of 100's is 1, that is 100 or C

the number of 1000's is 2,
that is 2000 or MM

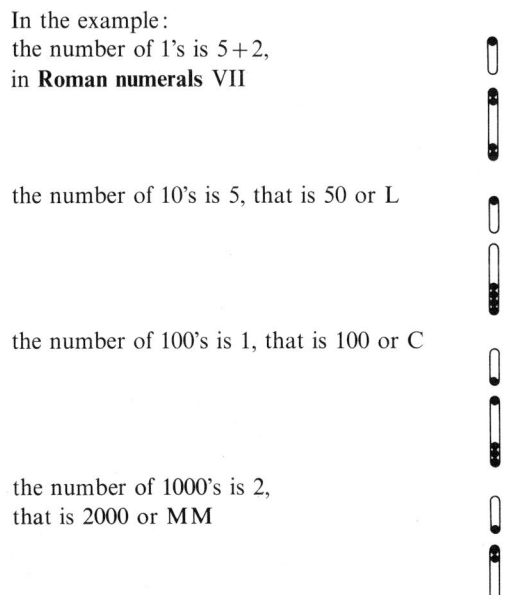

Thus the number is 2157 or MMCLVII.
The Romans had other types of abaci, particularly a table with **lines** drawn on it. Counters were then placed so as to denote numbers. If an abacus was not available calculations were done by drawing lines in the dust or sand and placing pebbles in the appropriate places. The Latin for pebble was **calculus** and from this we get the word calculate.

ROMAN SYSTEM
The main **numerals** are I (one), V (five), X (ten), L (fifty), C (one hundred) and M (one **thousand**). Repetition of a numeral indicates **addition** (XXX represents 30). A numeral placed after one of greater **value** indicates **addition** (XXI represents 21). A numeral placed before one of greater value indicates **subtraction** (IX represents 9).

LATIN *Roma*, Rome.
GREEK *systema*, to place together.

ROOT
1 The **solution** of an **equation**.
Example: $2x+6 = 12$. 3 is a root. $(x-2)(x-4) = 0$.
2 and 4 are roots of this equation.

2 See **square root** ($6 \times 6 = 36$. 6 is the square root of 36).
Cube root ($6 \times 6 \times 6 = 216$. 6 is the cube root of 216).
Fourth root ($6 \times 6 \times 6 \times 6 = 1296$. 6 is the fourth root of 1296).
Fifth root, sixth root and others follow a similar **pattern**.

ROPE STRETCHERS
In ancient Egypt rope stretchers made **right angles** that were required for building or **surveying**. A rope knotted into twelve **equal** sections was used to form the right angle. One man would hold the two ends together at the **point** (A) where the right angle was required. Another rope stretcher held the knot that was 3 sections from A and another held the knot that was 4 sections from A. By stretching the rope tight a right angle was formed at A.

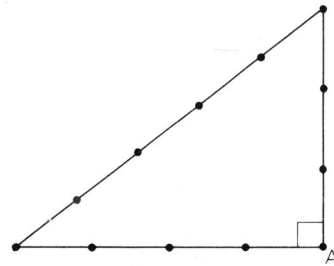

This can be proved correct by **Pythagoras Theorem**. This states that if the **square** on the longest **side** of a **triangle** is equal to the **sum** of the squares on the other two sides then the triangle is right-angled. The right angle is opposite the longest side. In the case illustrated $3^2 + 4^2 = 5^2$ ($16+9 = 25$) so there is a right angle at A.

ROTATION

Turning.

Example: A body may rotate about a fixed **point**.

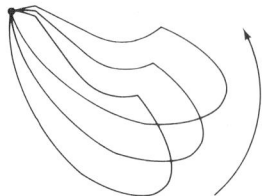

ROTATIONAL SYMMETRY

When **rotation** of a **figure** about a **point maps** the figure on to itself then the figure has rotational symmetry about that point. In simple **terms** a turn must leave the figure so it looks the same as before.

Examples:

The four vanes of the windmill have rotational symmetry about the **point** where they meet. An **equilateral triangle** has rotational symmetry about its **centre**, O.

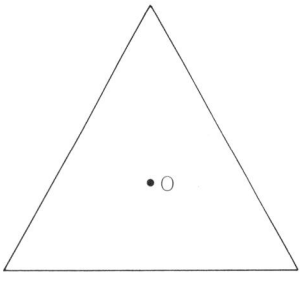

LATIN *rotatum*, to turn.
GREEK *syn*, together; *metron*, a **measure**.

ROUND

1 Curved, especially if approximating to all or part of **circle** or **sphere**.

2 To round a **number** we express it to a required degree of accuracy.
Example: 3847 rounded to the nearest 10 is 3850.
(See ROUNDED NUMBER.)

ROUNDED NUMBER

A **number** that is expressed with a more appropriate number of **digits**.
Example: The **length** of a road is given as 1223.89 **metres**. Due to inaccuracies in measuring this is probably only correct to the nearest metre so the length could be rounded to 1224 metres.

ROW

A **horizontal** arrangement of objects or **symbols**.
Examples:

```
× × × × × × × ×
× × × × × × × ×    3 rows
× × × × × × × ×
```

Vertically arranged objects or symbols are called **columns**. The above diagram has eight columns.
These chairs are arranged in rows.

OLD ENGLISH *raw*, row.

RULE

1 A procedure for carrying out a process.
Example: To find the **perimeter** of a **square multiply** the **length** of a **side** by 4.
2 To draw a **straight line**, as when using a **ruler**.

RULER

A graduated **scale** for measuring **length**.
A ruler can also be used for drawing **straight lines**.
(See STRAIGHT EDGE.)

RUSSIAN ABACUS

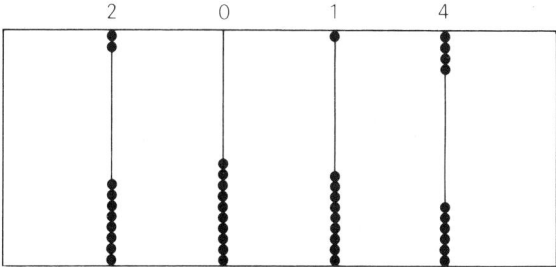

This version of the Russian abacus is called tsochottii and there are ten beads on each wire. Beads are pushed to the top to denote **numbers** in the usual **base** ten **notation**. See **denary system**. The illustration shows $2(1000)+0(100)+1(10)+4(1)$, that is 2014.
Another form (s'choty) has the wires horizontally. Two of the wires have only four beads on each, the others have ten. This form is useful for **calculations** involving money as the $\frac{1}{4}$ rouble and $\frac{1}{4}$ kopeck can be represented. The abacus is still widely used throughout Russia.

RUSSIAN MULTIPLICATION

A method of **multiplication** that only requires multiplication and **division** by 2 and **addition**.
Example:

To **evaluate** 34×18.

34 : 18	**Multiply** 34 by 2 and **divide** 18 by 2.
68 : 9*	Multiply 68 by 2 and divide 9 by 2.
	Ignore the **remainder**.
136 : 4	Multiply 136 by 2 and divide 4 by 2.
272 : 2	Multiply 272 by 2 and divide 2 by 2.
544 : 1*	

Mark the **odd numbers** in the **column** on the **right** (9 and 1).
Add the numbers on the left of those odd numbers (68 and 544). $34 \times 18 - 68 + 544 - 612$.
This method was said to have been invented in Russia to help people who were not able to multiply in the normal way but needed to work out questions such as the one above.

S

SAME

1 Not different. **Identical.**

Example: We have the same **number** of fingers on each hand.

2 **Equal** or very nearly equal. For example if two **lengths** are said to be the same they are actually compared or measured only to within certain **limits**. If we measure to the nearest **millimetre** then 28.3 mm and 28.4 mm are said to be the same, both being recorded as 28 mm.

3 Coming within one broad category or having one **attribute** in common. **Shapes** that all have three **sides** are the same in this respect and are all given the name **triangles**. They could differ in the **size** of the **angles** and the length of the sides.

SAMPLE

A relatively small **number** of people, **measurements** or **values** that are assumed to **represent** the **population** (all the ones from which the sample is selected). From the sample a reasonable forecast can then be made about the population.

For example by taking samples of the **sizes** of shoes worn by children at various ages a firm can ensure that they make the right **proportion** of shoes at each size.

(See RANDOM SAMPLE.)

SAND RECKONER

One of **Archimedes'** works on **finite numbers**. In it he showed how to **represent** the number of grains of sand that could be contained in a **sphere** that has its **centre** at the centre of the Earth and its **radius** the distance from that centre to the sun. See diagram at top of next column.

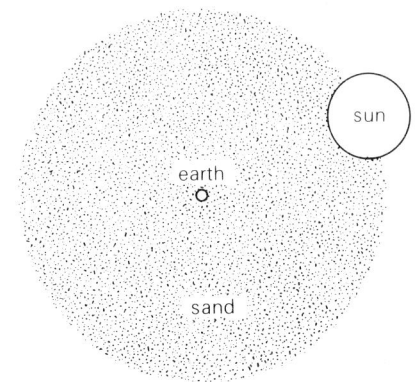

OLD ENGLISH *sand*, sand; *gerecenian*, to explain.

SATISFY

Fit stated conditions.

Example: Given the **equation** $2 \times \square = 8$, find $\square$. 4 will satisfy this equation since $2 \times 4 = 8$. 5 does not satisfy it since $2 \times 5 \neq 8$. $\neq$ means is not **equal** to.

SCALAR

A **quantity** that has **magnitude** but not **direction**. (Compare with VECTOR which has both magnitude and direction.)

Example: £3·20 and 9 **centimetres** are scalars. **Velocity** is a vector. (When direction is not given we use the **term speed**.)

SCALE

A measuring device frequently consisting of a **set** of **points** on a **line** as in the scale for a map.

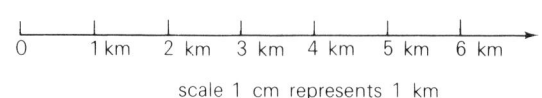

scale 1 cm represents 1 km

Scales can be seen on **thermometers**, **spring balances**, **graphs** and **slide rules**.

LATIN *scala*, a ladder.

SCALE DRAWING

A drawing, **plan** or **map** in which distances on the ground (or on another map) are represented by a **similar shape**. **Corresponding** distances are in **proportion**.

Example:
Any **length** in B is three **times** the corresponding length in A. (See SCALE.)

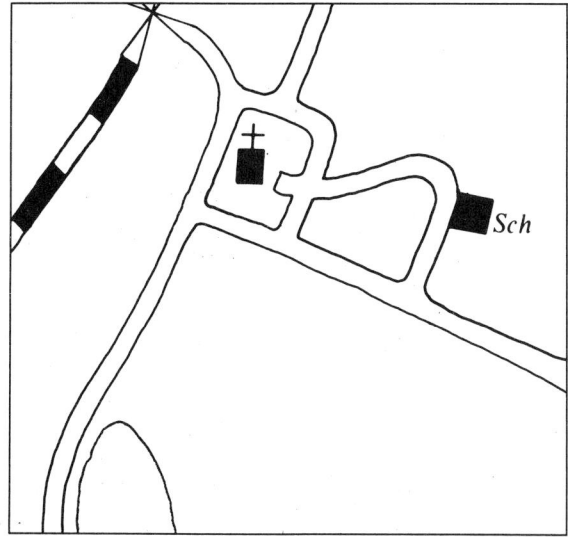

SCALENE TRIANGLE

A **triangle** with no two **sides equal**.

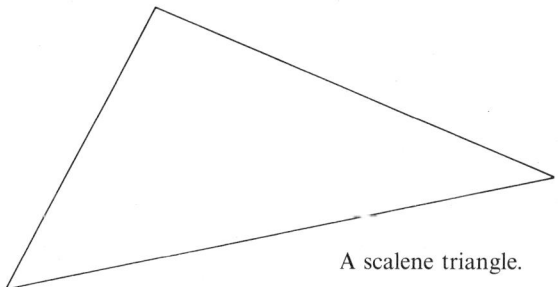

A scalene triangle.

GREEK *skalenos*, **uneven**.
LATIN *tri*, three; *angulus*, **angle**.

SCALES

A general **term** for any **instrument** used to find or compare **weights** or **masses**. It includes the simple **balance** consisting of two scale pans, the **spring balance** and many other forms.

SCORE

1 The **number** of **points** gained in a test or competition.

2 Twenty.
OLD ENGLISH *scoren*, sheared or cut. The score (1.) was kept by cutting notches in a piece of wood.

SCRATCH METHOD

A method of **division** used about 200 years ago. Also known as the galley method.

SECANT

A **line** that intersects (or cuts) a **curve** in one or more **points**. See INTERSECTION.
Example:

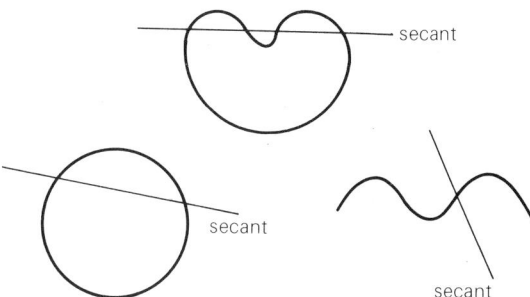

The **term** is especially used when referring to a line cutting a **circle**.

SECOND

1 A **measure** of **time**.
$\frac{1}{60}$ of a **minute**.
Very **accurate measurements** may be required in science. The second is then based on the radiation in a particular type of atom.

2 Second (2nd) comes after **first** when placed in **order**.

SECTION

See CROSS-SECTION and GOLDEN SECTION.
LATIN *sectum*, to cut.

SECTOR

A **region** bounded by two **radii** of a **circle** and the **arc** joining their end **points**. Two possible sectors are formed the smaller being the **minor** sector and the larger the **major** sector.

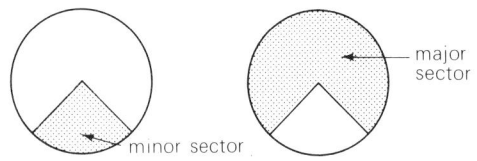

LATIN *sectum*, to cut.

SEGMENT

1 That part of a **line**, between two **points**.
(Note: a line is regarded to have **infinite length**.)

A B is a **line segment**.
It is usual to call line segments lines but this is not strictly correct.

2 The **region** bounded by an **arc** and the **chord** joining the two end **points** of that arc.

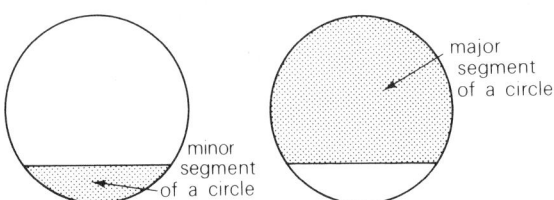

The smaller region of the circle is called the **minor** segment and the larger region is called the **major** segment.
LATIN *segmentum*, cut off.

SELLING PRICE

The **amount** that an article is sold for.
Example: A shopkeeper buys a coat for £30 from a manufacturer and sells it to a customer for £40.
Shopkeeper's **cost price** (C.P.) = £30.
Shopkeeper's **selling price** (S.P.) = £40.
Profit = S.P. − C.P. = £40 − £30 = £10.

SEMICIRCLE

Half of a circle. All **diameters** separate a circle into two semicircles.

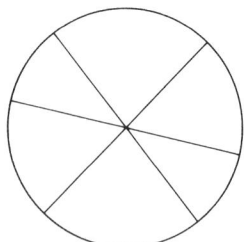

Note that when circle is taken to mean a curved **line**, a semicircle is an **arc** such as ⌢. If instead circle is taken to mean the inner **region** ⊘ then semicircle is a region such as ⌓.

SENSE

1 Given any **line** there are two possible **directions**, each is called a sense of direction. One is the **positive** sense and the other (or opposite) as the **negative** sense.
Example:

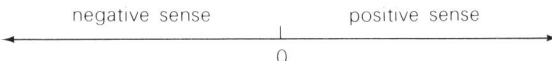

On a **number line** or **horizontal axis** of a **graph** the direction towards the **right** is usually taken as the positive sense (+ve) and the direction to the left as the negative sense (−ve).

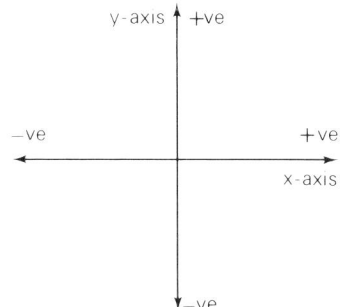

2 Also used in direction of **rotation**. **Anti-clockwise** is generally taken as the positive sense of rotation +ve↺ and **clockwise** as the negative sense −ve↻.

SENTENCE

A **statement**. In **mathematics** it may contain **numerals** and other **symbols**. A sentence may be **true** or false and open or closed.
Examples:
$2x = 10$ is a true sentence if $x = 5$. For other **values** of x it is false.
$3 \times \square = 27$ is an **open sentence** until the **variable** (in this case $\square$) is replaced by a **number**. It then becomes a **closed sentence**.

$3 \times \boxed{9} = 27$ is a closed sentence that is true.
$3 \times \boxed{11} = 27$ is a closed sentence that is false.

SEQUENCE

A **set** of **numerals** written one after the other such that there is a definite **relation** between one **term** and the next.
Example: 2, 6, 18, 54, 162, ...

Any **number** after the first is three **times** the preceding number. Sequence is sometimes used for any set of numbers written one after the other even when there is no relation between them.
Example: 11, 17, 8, 53, 19.

SERIES

The **sum** of the **terms** of a **sequence**.
Example: 2+4+6+8+10+12+14 is a **finite** series.
2+4+6+8+10+12+14 ... is an **infinite** series.

SET

Stated simply a set is a **collection** of objects (called **members** or **elements**) that are classed together so that it is known whether any object does, or does not, belong to that set.
Example: A set of **triangles**. Given any **shape** we can tell whether or not it belongs to the set of triangles. The set of clothes you are wearing at the **time** of reading this.

The **elements** or **members** of a set are written in curly **brackets** or **braces** { }. Set of vowels in the English alphabet = {a, e, i, o, u}. (The **order** of elements in a set does not matter so this is the same set as {e, i, a, u, o}.)

SET SQUARE

See picture opposite.

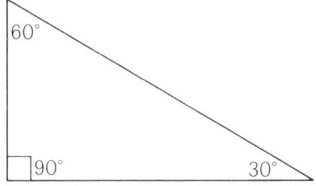 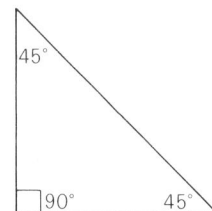

Instruments used for drawing **parallel lines**, **perpendicular** lines or particular **angles** (30°, 60°, 45°, 90°). A set square may have angles 60°, 30°, 90° or else 45°, 45°, 90°.
Example: To draw a line through C that is parallel to A B.

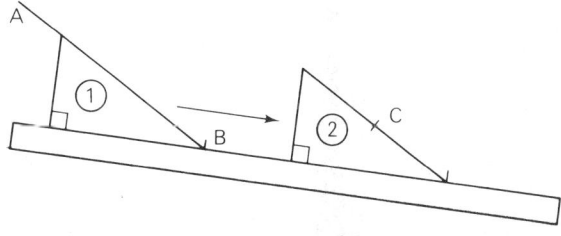

Method: Place one **edge** of the set square along part of A B ①. Place a **ruler** along one of the other edges such that the set square can be moved in the **direction** of the ruler to position ②. Draw the line on the **hypotenuse**.

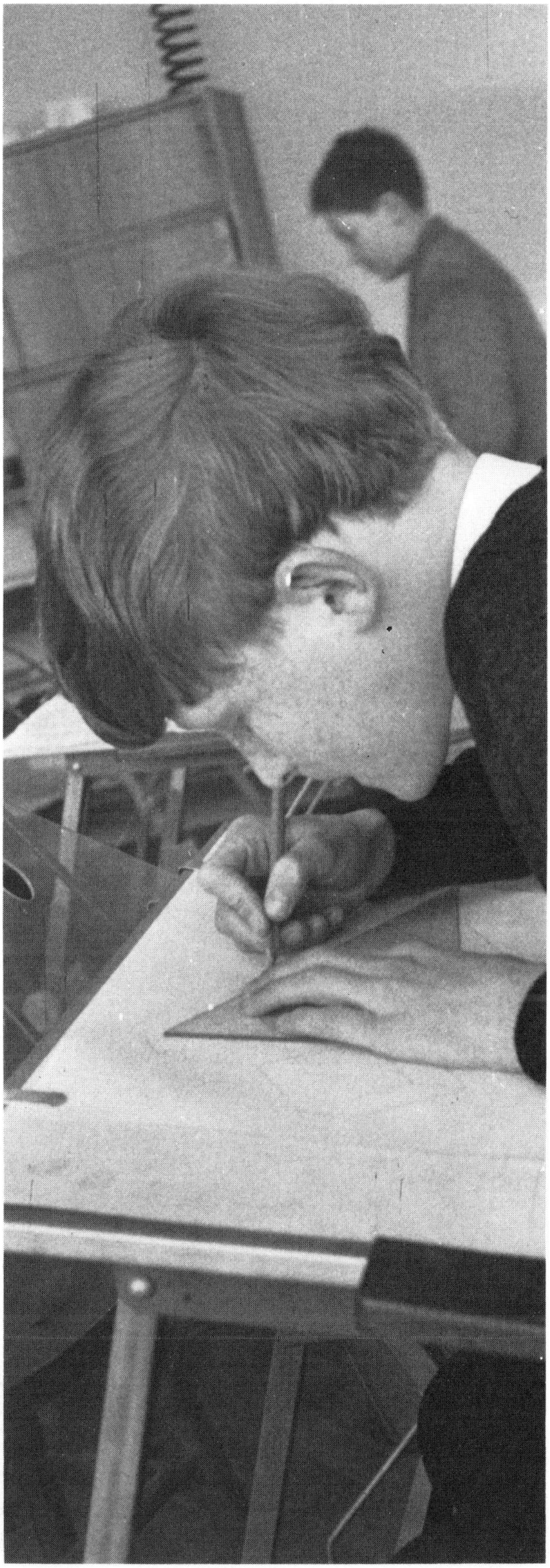

SEXAGESIMAL SYSTEM

A **number** system with sixty as its **base**.
(See BABYLONIAN SYSTEM.)

SHADOW RECKONING

1 A method of measuring **heights** by means of the sun's shadow.

Example: To find the height of a tree CB: A stick of known height is placed vertically at ED. AD and AB, the **length** of the sun's shadows are measured.

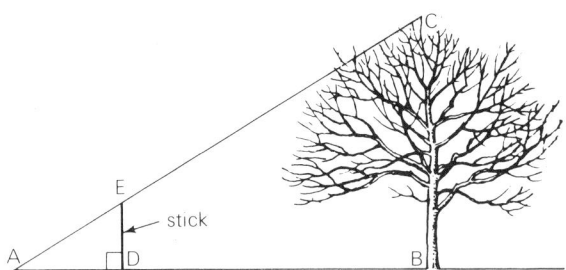

As **triangles** EDA and CBA are **similar** in **shape** $\frac{CB}{AB} = \frac{ED}{AD}$. These are all known except CB which can therefore be calculated. Suppose the lengths are ED 2 **metres**, AD 8 metres and AB 56 metres.

$$\frac{CB}{56} = \frac{2}{8} \qquad CB = \frac{2}{8} \times {}^{56} = 14$$

The tree is 14 metres high.

2 A means of telling the **time** by measuring the length of shadows. This method was useful in countries where there was a great deal of sunshine but has little value where the climate is cloudy. See picture below.

OLD ENGLISH *sceadu*, shade.
GERMAN *rechnen*, reckon.

SHAPE

The form. The way in which material is arranged to form an object. When the arrangement occurs frequently we give a special name to that shape as in **square**, **cuboid** or **cylinder**.

SHARE, SHARING

1 One of the two aspects of **division**, the other being **quotition**.

Example: Share 12 sweets between 4 people. They each get 3 sweets. $12 \div 4 = 3$.

An example of quotition would be: How many **groups** of 4 sweets can be taken from 12 sweets. The answer is 3 groups. We again write $12 \div 4 = 3$.

2 A Company may raise money by selling shares. These give the purchaser a 'share' or part of the company and its **profits**.

SHOPKEEPER METHOD

Although called a method of **subtraction** it is actually **addition**.

Example: A woman buys some apples for 38p and gives the shopkeeper a 50p coin.

The shopkeeper gives the customer the apples (38p) and then **adds** money to this until 50p is reached.

He might say 38p and 10p makes 48p, and 2p makes 50p.

OLD ENGLISH *sceoppa*, a treasury; *cepan*, keep, to store up.
GREEK *methodus*, method or way.

SHORT DIVISION

A method of **division** that is suitable when dividing by a small **number**.

Example:
$$4 \overline{)1\,3^1\!4^2\!8}$$
$$\underline{3\;3\;7}$$

The picture shows a primitive shadow **clock** from the upper reaches of the Nile.

This method is more usually laid out in a **similar** way to **long division**:

$$4)\overline{13\,3^14^28}\ .\qquad 337$$

SIDE
A **line** forming the **boundary** of a **figure**.
Example:

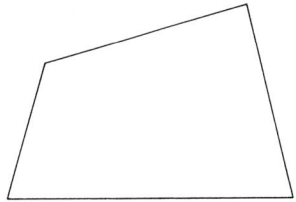

A **quadrilateral** has four **sides**.

SIEVE OF ERATOSTHENES
A method used by the Greek scholar Eratosthenes to find **prime numbers less than** a stated number. Method: Write down all the **natural numbers** from 2 up to, and including, the stated number. Suppose this is 27. Leave 2 but cross out every **multiple** of 2, that is 4, 6, 8, 10 and so on. 2, 3, 4, 5, 6, 7, 8, 9, 10, 11, 12, 13, 14, 15, 16, 17, 18, 19, 20, 21, 22, 23, 24, 25, 26, 27. Leave the next number 3, but cross out all the multiples of 3, that is 6, 9, 12, 15, 18, . . . Continue this with the next remaining number, 5 (4 has been crossed out) and so on. The numbers left are prime.
OLD ENGLISH *sife*, a sieve.

SIGN
A **symbol** used to denote an **operation**.
Addition sign $+$.   **Subtraction** sign $-$.
Multiplication sign $\times$.   **Division** sign $\div$.
Square root sign $\sqrt{\ }$.

SIGNED NUMBER
A **negative number**, **positive number** or **zero**. Zero is included as it can be regarded as $+0$ or -0. Signed numbers are also called **directed numbers**.

SIGNIFICANT FIGURES
The **figures** or **digits** in a **number** that are acceptable as an **approximation**.
318.7 to 3 significant figures (3 sig. fig.) is 319.
0.1034 to 3 sig. fig. is 0.103.
(Note the 0 in the **units position** is not a significant figure but the 0 in the second place of **decimals** is).
It would not be sensible to give the **length** of a room to the nearest **millimetre** and a measurement such as 10.3894 **metres** would therefore be approximated to

↑
Number of
millimetres
10.39 m using 4 sig. fig.

SIMILAR FIGURES
Figures in which **corresponding angles** are **equal** and corresponding **sides** are in **proportion**.

The solid shapes are similar.

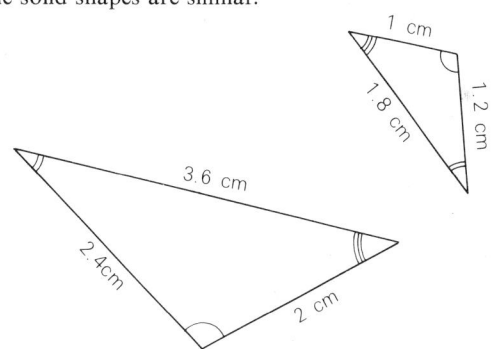

These two **triangles** are similar.

SIMPLE FRACTION
Also called **common fraction** and **vulgar fraction**.

SIMPLE INTEREST
Interest in which the money upon which interest is paid remains the same throughout the period of the loan.
Example: £100 invested at 10 **per cent** simple interest is worth £100 + £10 at the end of the **year**. During the second year interest is again given on the £100 and *not* on £110. The **total** interest is therefore £20.
(See COMPOUND INTEREST.)

SIMPLE PRACTICE
See PRACTICE.

SIMPLE PROPORTION
See DIRECT PROPORTION.

SIMPLIFY
Write in a shorter form.
Example: Simplify: $3+(4\times2)-5+(8\div2) = 3+8-5 +4 = 10$.

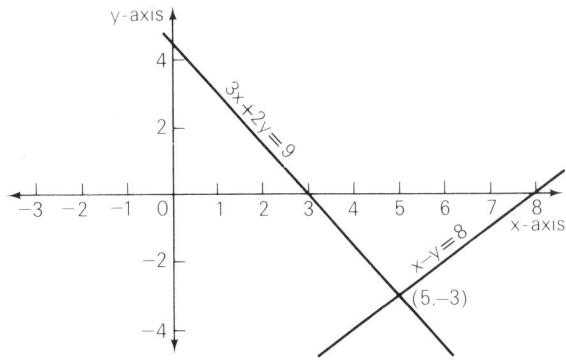

SIMULTANEOUS EQUATIONS

Two or more **equations** that have a common **solution**.

Example: (i) $3x + 2y = 9$
   (ii) $x - y = 8$ (Multiply (ii) by 2)
   (iii) $2x - 2y = 16$

Add (i) $5x = 25$
and (iii) $x = 5$

Substitute in (ii) $5 - y = 8$, $5 - 8 = y$, $y = -3$
Solution $x = 5$, $y = -3$
See diagram above.

LATIN *simul*, at the same **time**; *aequalis*, **equal**.

S.I. SYSTEM

See SYSTÈME INTERNATIONALE D'UNITÉS(S.I.).

SIZE

The **dimensions** or **magnitude**. The **amount**.

SKELETONS

Figures formed by the **edges** of **solids**. Skeletons can be made from straws and pipe cleaners.

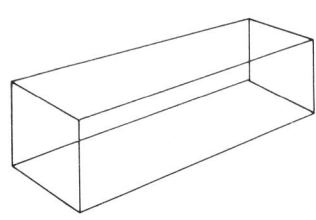

Skeleton **cuboid**

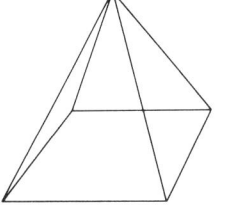

Skeleton **pyramid**

GREEK *skeleton*, dried up or withered. A body.

SKEW LINES

Lines that are not **parallel** but do not intersect. Skew lines do not lie in the same **plane**.

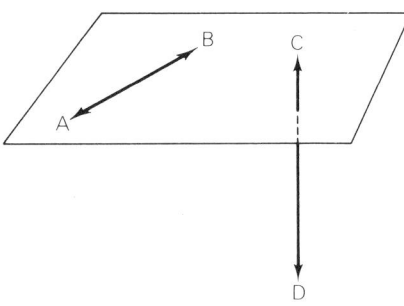

In this **cuboid** AB and CD are skew lines.

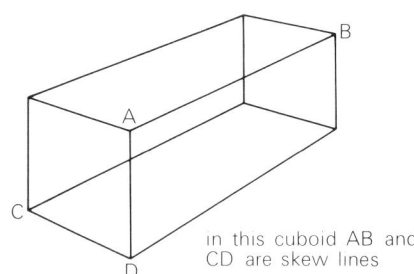

in this cuboid AB and CD are skew lines

OLD FRENCH *eskiuer*, led to.
MIDDLE ENGLISH *skewen*, slanting, twisted.
OLD ENGLISH *lin*, flax. (See line.)

SLANT HEIGHT

1 For a **cone**: The distance from the **vertex** to a **point** on the rim of the **base**.

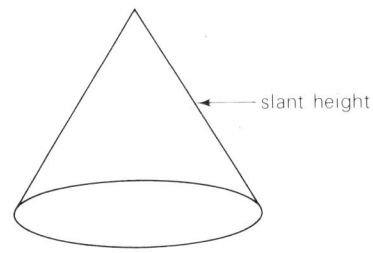

slant height

2 For a **pyramid** with a base that is a **regular polygon**: The distance from the vertex to the midpoint of any **line** forming the base.

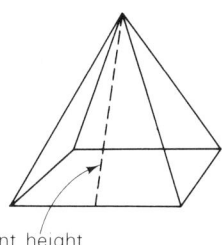

slant height

MIDDLE ENGLISH *slent*, slope.
OLD ENGLISH *heah*, high.

SLIDE
To slip or glide.

SLIDE RULE
A calculating device used for **multiplication, division** and other mathematical **operations**. Basically it consists of two logarithmic **scales** (see LOGARITHM) that enable multiplication and division to be replaced by **addition** and **subtraction**.
OLD ENGLISH *slidan*, to slide.
OLD FRENCH *reule*, to rule.

SLOPE
Gradient. The **rate** at which a **curve** rises or falls. The slope of a **line** is measured by the **vertical** rise (or fall) divided by change in the **horizontal** distance.

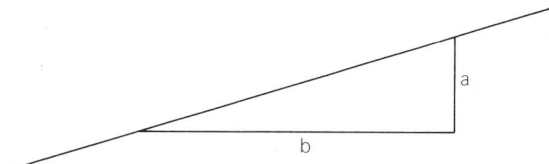

The slope or gradient is $\dfrac{a}{b}$.

The slope of a **curve** is the slope of the **tangent** at any given **point**.

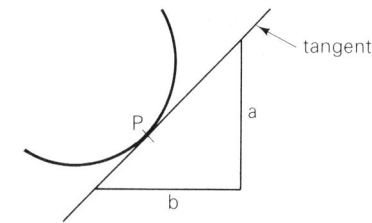

The slope of the curve at P is $\dfrac{a}{b}$.

OLD ENGLISH *slupan*, slope.

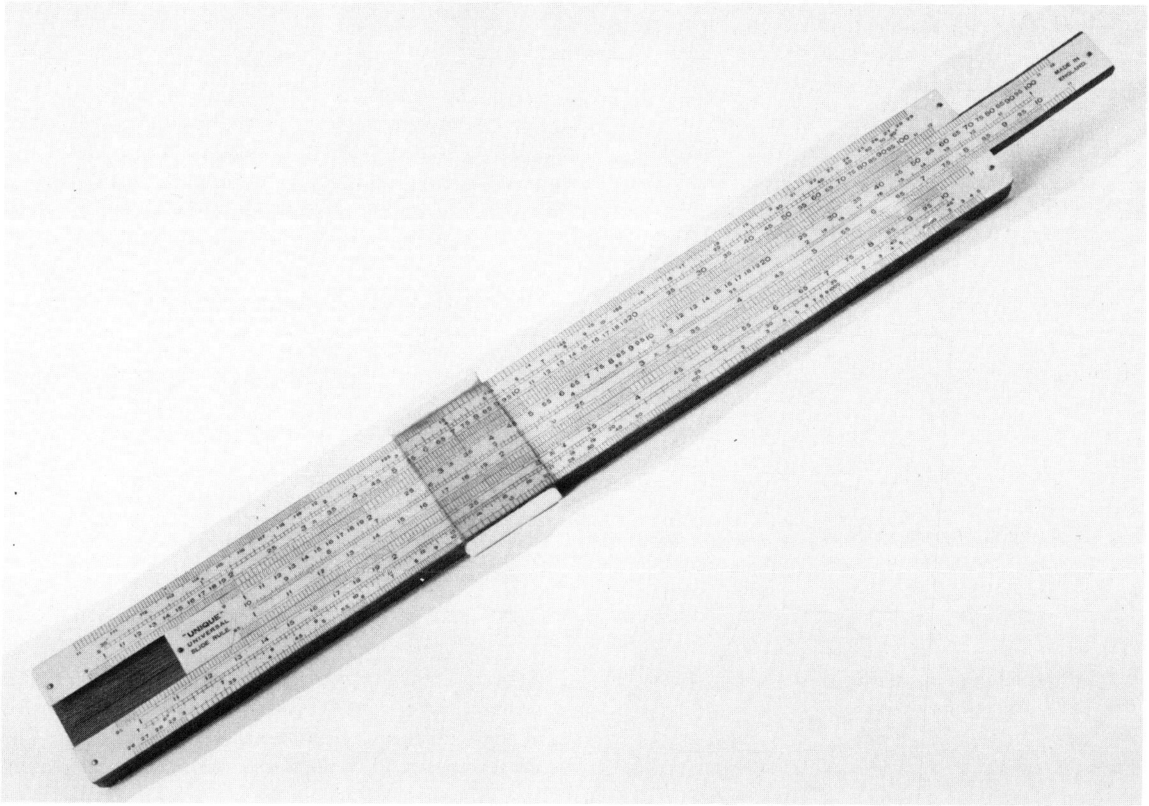

SLOW

1 Taking a longer **time** than is necessary. Slow is a relative **term**, a motorist would be moving slowly but, to a person running, that **speed** would be **fast**.

2 A **clock** or watch is slow when it shows a time earlier than the correct time.
(See FAST.)

SMALLER THAN   See LESS THAN.

SOLID

A **figure** with three **dimensions**.
Example: A **sphere**, **cube**, book or pencil. The figure may only be a **skeleton**. For example the twelve **lines** forming the skeleton of this **cuboid** from a solid.

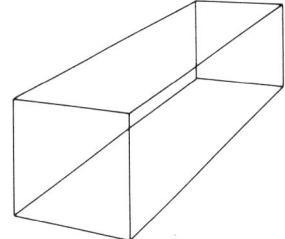

SOLID GEOMETRY

The study of **three-dimensional shapes** and their **properties**.
LATIN *solidus*, **solid**.
GREEK *ge*, earth; *metron*, **measure**.

SOLIDUS

The slanting **line** separating the **numerator** from the **denominator** of a **fraction**.
Example: 3/4   7/8
              three quarters   seven eighths.

(The **term** solidus is sometimes used when referring to the **bar** separating the numerator and denominator of a fraction. $\frac{3}{4} \rightarrow$ bar or solidus).

SOLUTION

The answer to a **problem**.
$2x + 3 = 9$ has a solution $x = 3$.
The **set** of **values** for a **variable** that make a **statement** true is called the solution set. In the above example the solution set is {3}.

SOLVE

To find the **solution** or answer.

SOMA CUBE

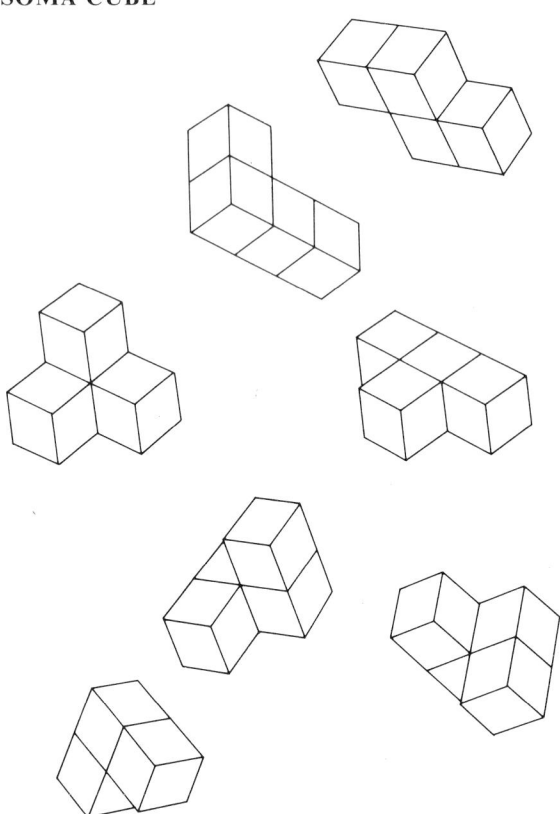

The seven **shapes** above can be fitted together to make a **cube** of **sides** three **units**. This is called a soma cube. It was invented by the Danish poet and mathematician Piet Hein. Many other interesting shapes can be made. Soma was a drug that Aldous Huxley wrote about in The Brave New World. It was invented to fill in spare **time** pleasantly. The soma cube puzzle can also do this.

SOROBAN

A Japanese form of an **abacus**.

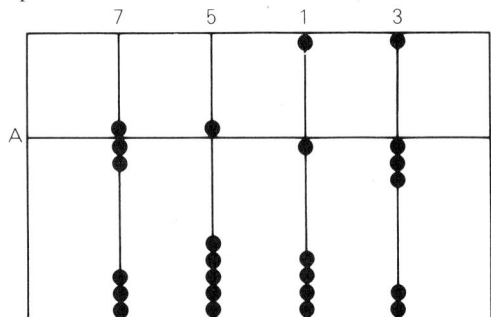

The beads in the top part of the frame **represent** five if they touch the bar marked A. Beads in the lower part represent 1 if they are in the **position** nearest to the bar A. The soroban shows 7513. Japanese word for a calculating board.

SPACE
A **three-dimensional region**.
A **set** of all the **points** in such a region.

SPAN
The distance from the thumb to the tip of the little finger of a stretched hand.

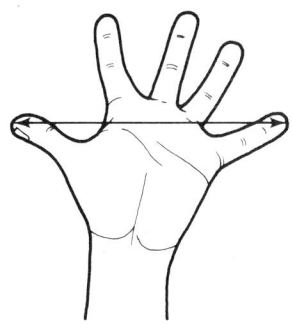

2 spans are approximately one **cubit**.

SPECIFIC GRAVITY
The **ratio** of the **weight** of a substance to the weight of an **equal volume** of water.

$$\text{Specific gravity} = \frac{\text{Weight of substance}}{\text{Weight of equal volume of water}}.$$

SPEED
The distance travelled in a **unit** of **time**.
Examples: The speed of light is approximately 300 000 **kilometres** per **second**.
A car covers 158 kilometres in 2 **hours**. Its **average** speed is 79 kilometres per hour.

SPHERE
1 The **surface** consisting of all **points** that are a given distance from a fixed point (the **centre**).

2 The **solid** bounded by the surface in 1 (above) together with the **region** inside that surface.

SPIRAL
A **curve** which is the path of a **point** that moves from a fixed point, so that its distance from that fixed point is always increasing.

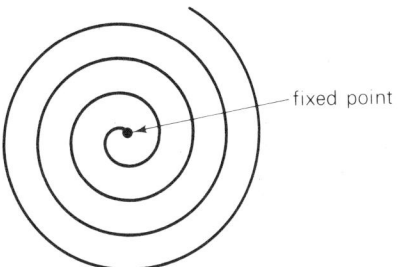

There are many different types of spirals, such as **Archimedes'** spiral, **Pythagoras'** spiral, **equiangular** spiral and parabolic spiral.

SPIRIT LEVEL
An **instrument** for determining whether or not a **surface** is **horizontal**.
The liquid used is a spirit and a bubble is trapped in it. When the bubble is in the **centre** the spirit level is horizontal and so is the surface on which it stands.

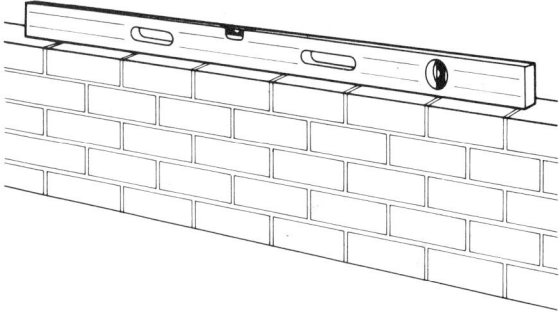

SPRING
A coil of wire in the form of a **helix**.

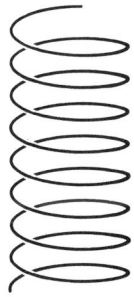

When a **force** is applied the spring is compressed so that, with a suitable **scale**, it can be used to **measure** forces. See also SPRING BALANCE. Energy is stored in a compressed spring and it can be released gradually as when a spring is used to turn the **hands** of a **clock**.

SPRING BALANCE
A **balance** in which the **force** exerted by the **weight** of an object is balanced by the force in a compressed or extended **spring**. The weight of the object can be read directly from a suitable **scale** on the balance.

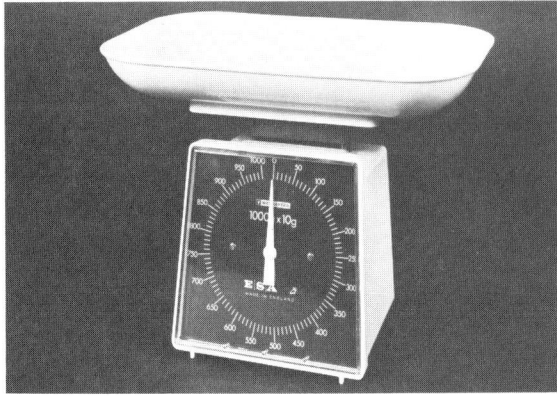

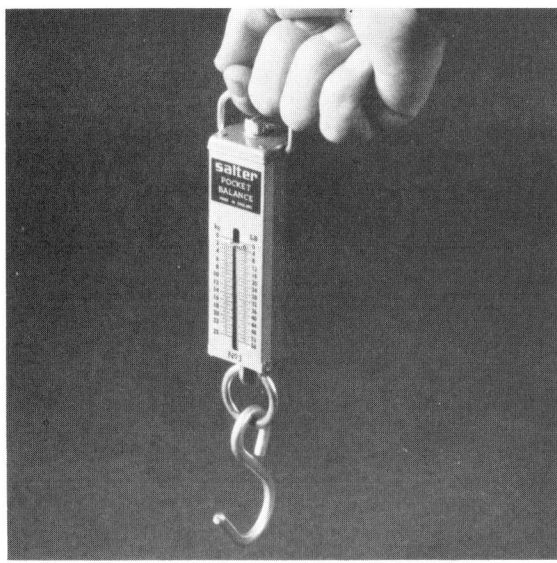

SQUARE
Three definitions are given.

1 A **quadrilateral** with four **equal sides** and four **right angles.**
(Strictly speaking it is enough to say that it has four equal sides and one right angle. From this it can be proved that the other **angles** are right angles.)

2 A **parallelogram** with two adjacent sides equal and a right angle.
(The fact that all the sides are equal and all the angles are right angles can be deduced from this information.)

3 A **rhombus** with one of its angles a right angle.
(It can then be proved the other angles are right angles.)
FRENCH *esquarre*, square.

SQUARE MEASURE
Units in which **areas** are measured and their **relations** to one another.
Example: Square **centimetre** (cm^2), square **metre** (m^2).
The relation between units are given in **tables**:

$100\,mm^2 = 1\,cm^2$ $100\,cm^2 = 1\,dm^2$
$100\,dm^2 = 1\,m^2$ $100\,m^2 = 1\,dam^2$
$100\,dam^2 = 1\,hm^2$ $100\,hm^2 = 1\,km^2$

SQUARE NUMBER
A **number** that can be expressed as the **product** of two **equal** numbers.
Example: $25 = 5 \times 5$ so 25 is a square number.
Square numbers can be represented by **dots** arranged to form a **square**.

Examples:

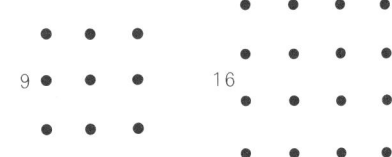

SQUARE OF A NUMBER
A **number** multiplied by itself. $3 \times 3 = 9$ so the square of 3 is 9. This can also be written as $3^2 = 9$ (read as three squared **equals** nine).

SQUARE PAPER
Any paper ruled in **squares**. This is convenient for plotting **graphs** and also for tabulating results.

SQUARE ROOT
The square root of a **number** is that number which multiplied by itself gives the required number.
Example: What is the square root of 64?
Since $8 \times 8 = 64$, 8 is the square root of 64.

STANDARD UNITS
Units that are accepted throughout a country, or **group** of countries.
Example: The **metre** is the standard unit of **length** in the **metric system**.

STATEMENT
A mathematical **sentence**, such as
$5x + 3 = 8$ or $\triangle = 2\square + 6$.
A statement may be **true** or false according to the **value** that is substituted for the **variable**. $5x + 3 = 8$ is true if $x = 1$ but false for all other values of x.

STATICS
The study of the **forces** on bodies or particles that are at rest.

STATISTICS

The **collection**, **representation** and **analysis** of information. (The information is called **data**.)

LATIN *status*, state. Originally statistics was information useful to the country or state.

STATUTE MILE

Also called a land **mile**.

A **measure** of distance in the **imperial system**. 1 statute mile **equals** 5280 **feet**, or 1760 **yards** or approximately 1.609 **kilometres**.

For ease of conversion 5 miles is approximately 8 kilometres. Mile is derived from mille which is Latin for a **thousand**. The Roman mile was 1000 **paces** (Latin-passus). These were **double** paces, of about 5 feet each, so the mile was approximately 5000 feet. This was changed to 5280 feet so as to make land measuring easier. The **furlong** was the main measure in use and 8 furlongs equal 5280 feet (8 × 660 feet).

From the **year** 1500 onwards most people accepted 5280 feet as a mile but this was not approved by law until 75 years later.

LATIN *statutum*, that which has been set up (in this case by law); **mille**, a thousand.

STEELYARD

A **machine** for weighing. It consists of a **lever** with a short arm for the object being weighed and a long arm on which a small **weight** can be moved. The long arm contains a **scale** on which the weight of the object can be read.

STEEP

A considerable rise or fall especially of a road.
Examples: A danger sign warned motorists of the steep hill. There was a steep rise in the cost of houses.

STERN MATERIAL

Coloured **cubes** and **rods** used by young children to develop their ideas of **number**. These and other related material were developed by Dr. Catherine Stern in America.

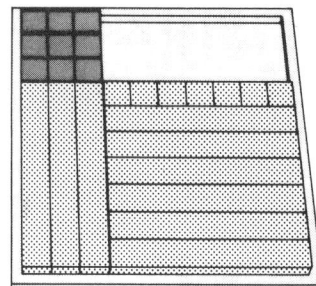

STOCK TAKING

The checking of stock in a shop.

The **amount** sold is the **difference** between that held at the last stock taking and that now held. This **value** of the amount can be checked against the cash received from sales.

STONE

A **measure** of **mass** or of **weight** in the **imperial system**. 1 stone = 14 **pounds**.

STOP WATCH

A watch that can be started and stopped by pressing a button, or moving a **lever**.

It is used for races and events that must be timed accurately.

STRAIGHT

Not curved. Not having any bends.

STRAIGHT ANGLE

An **angle** of 180° or two **right angles**.

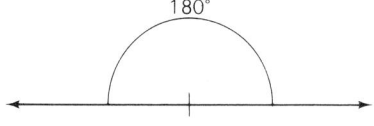

STRAIGHT-EDGE

An **instrument** for drawing **straight lines.** It differs from a **ruler** in that it does not have a **scale** for measuring.

STRAIGHT LINE

A **line** extends infinitely in both **directions**. This is shown by arrows.

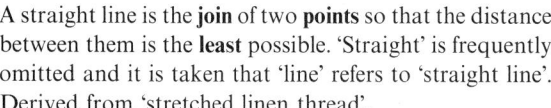

A straight line is the **join** of two **points** so that the distance between them is the **least** possible. 'Straight' is frequently omitted and it is taken that 'line' refers to 'straight line'. Derived from 'stretched linen thread'.

STRUCTURE

The basic **elements, concepts, relations** and principles that apply to a mathematical system.

SUAN-PAN

Also spelt swanpan. A Chinese **abacus.**

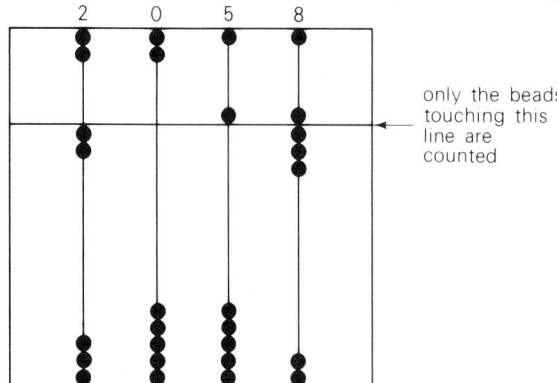

Each top bead **counts** as 5, the ones in the bottom part 1.
CHINESE *suan-pan*, a board for calculating.

SUBSET

If every **element** (or **member**) of **set** A is also an element of set B then set A is a subset of set B.
Examples: The boys in a class of boys and girls form a subset of the set of children in the class.
Even numbers form a subset of the set of **whole numbers**.

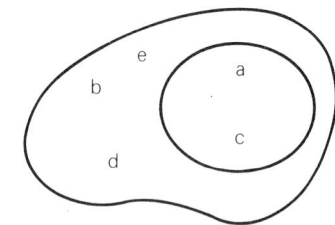

{a, c} is a subset of the set {a, b, c, d, e}.
The set {1, 2, 3} is regarded as a subset of {1, 2, 3}. All All subsets except the set itself are called **proper subsets**.

SUBSTITUTE

To replace a **variable** by a **number**.
Example: In $6x + 8$ we can substitute 5 for x. The expression then has the **value** 38.

SUBTEND

To be opposite to.
Arc AB is said to subtend the **angle** ACB.

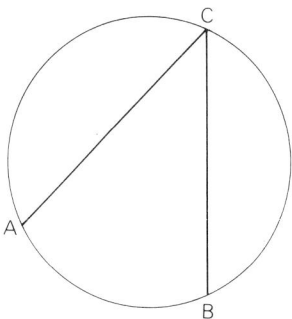

SUBTRACT, SUBTRACTION

There are three interpretations.
1 The **inverse** of **addition**. $14 - 8$ is thought of as 'what must be added to 8 to give 14?'

2 **Comparison.** Two **numbers** are compared in **magnitude**.
Example: $14 - 8 = 6$. 14 is seen to be **larger than** 8, the **difference** being 6.

3 **Take away.** 8 is taken away from 14. $14 - 8 = 6$.

SUBTRAHEND

A **number** that is subtracted from another number.
Example: 43 is the **minuend**.
  -18 is the **subtrahend**.
  25

LATIN *subtrahendus*, requiring to be removed or drawn away.

SUCCESSION

Following one after the other in a definite **order**.
A **term** that follows another is called its successor.

SUM

1 The **result** of adding two or more **numbers** or **quantities**.
Examples: $28 + 93 = 121.$ $17\,\text{cm} + 19\,\text{cm} = 36\,\text{cm}.$
$27° + 43° = 70°.$ The sums are 121, 36 cm and 70°.

2 The **operation** of **addition**.
Example: Sum these numbers 8, 7, 3.

3 Incorrectly used for any **arithmetic calculations** involving **addition, subtraction, multiplication** or **division**.
Examples: $32 - 19,\ 218 \times 3,\ 124 \div 4.$

SUMMIT

The highest **point** of a hill or mountain.

SUPPLEMENTARY ANGLES

Two **angles** whose **sum** is 180°.
Example: 83° and 97° are supplementary. 83° is called the supplement of 97° and 97° is the supplement of 83°.

SURFACE

A **set** of **points** making a space in **two dimensions**. They may form a **plane** (a flat surface) or be curved, as is the surface of a **sphere**.

SURVEYING

The mapping of a piece of land. There are many techniques such as a **plane table** surveying and **triangulation**.

SYMBOL

A letter, **numeral** or other mark that **represents** a **number**, **operation** or any mathematical idea.
Examples: x (a **variable**).
  π (the **ratio** of the **circumference** to the **dia-** to the **diameter** for any **circle**).
  IV (Roman numerical for four).

SYMMETRIC DIFFERENCE

The **set** of **elements** that are in one or other of two given sets but not in both.
Example:

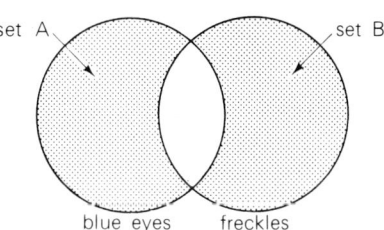

set A   set B
blue eyes   freckles

The shaded **area represents** the symmetric difference of sets A and B.
It is written as $A \triangle B$.
Set A is children with blue eyes and set B is children with freckles. $A \triangle B$ is the children who are blue-eyed but do not have freckles, together with these who have freckles but do not have blue eyes.

SYMMETRIC PROPERTY OF RELATIONS

A **relation** such that if a is related to b then b is related to a in the same way.
Examples: John is married to Mary, hence Mary is married to John. ('Married to' is a symmetric relation.)
AB is **parallel** to CD. Therefore CD is parallel to AB. (The relation 'is parallel to' is symmetric.)
Jean is fatter than John. It cannot be true that John is fatter than Jean. (The relation 'is fatter than' is *not* symmetric.)

SYMMETRY

The **correspondence** of **points** that are on opposite **sides**, and **equal** distances from a point (see **point symmetry**), **line** (see **bilateral symmetry**) or a **plane** (symmetry about a plane). Also see **rotational symmetry**.
Examples:

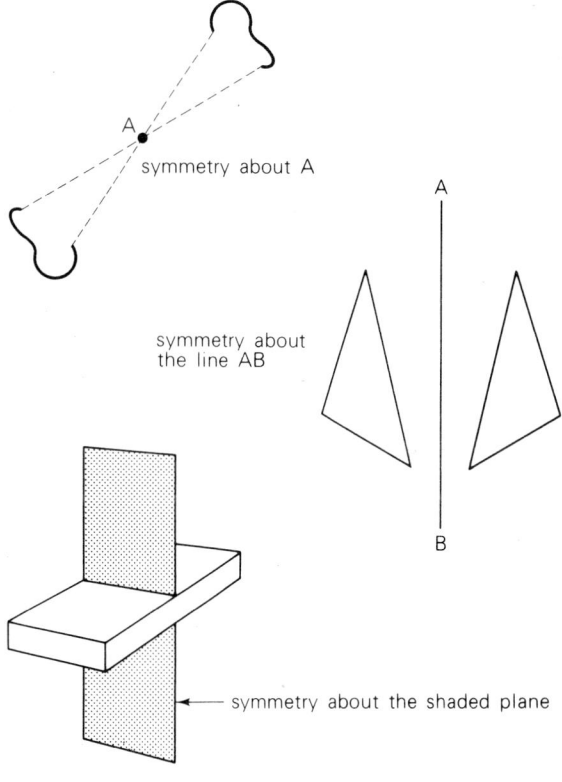

symmetry about A

symmetry about the line AB

symmetry about the shaded plane

SYSTÈME INTERNATIONALE D'UNITÉS(S.I.)

Literally this means the international system of **units**. This system was adopted in 1960 by the General Conference of **Weights** and **Measures**. It is based on the **metre (m) kilogram** (kg), **second** (s), **ampere** (A) (for electric current) and several other more specialised units. Prefixes are used to express **multiples** and parts of the basic units. The main ones are kilo (1000), **hecto** (100), **deca** (10), **deci** ($\frac{1}{10}$), **centi** ($\frac{1}{100}$) and **milli** ($\frac{1}{1000}$).

T

TABLE(S)

1 An orderly arrangement of **numbers** or letters, usually in **rows** and **columns**.

2 An arrangement of **multiplication** facts in **order**, called the multiplication tables.

Example:

$8 \times 1 = 8$
$8 \times 2 = 16$
$8 \times 3 = 24$
$8 \times 4 = 32$
etc.

The table of 8's.

Note: 8×4 is read as 8 multiplied by 4 and not as 8 **times** 4. Thus the given table is the 'table of 8s' and not the '8 times table'.

LATIN *tabula*, a board.

TABULATE

To arrange in the form of a **table**.

TAKE AWAY

One interpretation of **subtraction** when objects are considered to be removed or taken away.

Example: A girl has 16 sweets and eats 7 of them. She still has 9. $16 - 7 = 9$.

Subtraction can also be considered as the **inverse** of **addition** and as **comparison**.

TALLY STICK

A stick on which notches were cut to **represent** each object to be counted. Tally sticks were also used to keep **accounts**. The stick was cut down the middle so that one **half** could be kept as a receipt by the person paying.

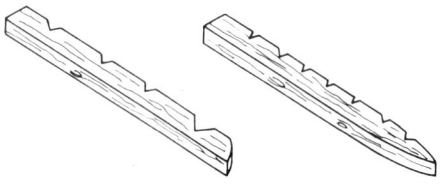

TANGENT

1 A **straight line** that **touches** a **curve** at one **point** only.

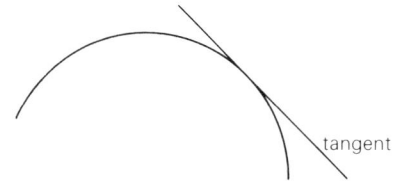

(The straight line, if produced, must not cut the curve. The line shown below is *not* a tangent.)

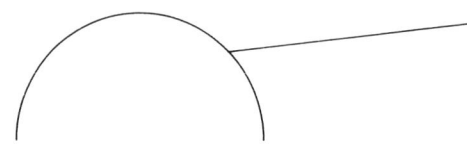

2 A **plane** that meets a curved **surface** at one point only or meets the surface along a **line**, as when a **flat** surface is placed on the curved surface of a **cylinder**.

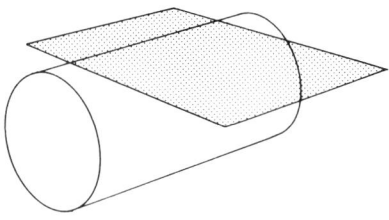

Tangent plane to a cylinder.

3 A **ratio** in **trigonometry**.

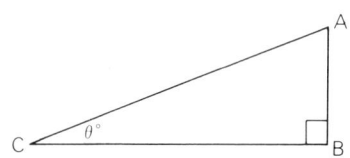

The tangent of **angle** $\theta°$ is the ratio $\dfrac{AB}{CB}$.

LATIN *tangens*, touch.

TANGRAM

A Chinese puzzle invented about 4000 **years** ago. A **square** is cut into seven pieces as shown. Many interesting **shapes** can then be made from the pieces.

There are two main forms of the puzzle:

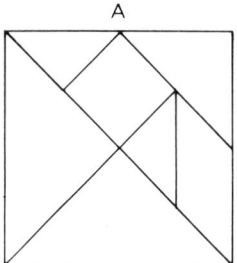 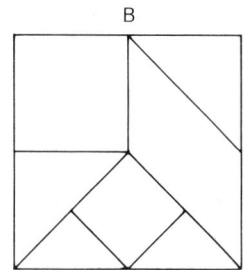

A   B

These pictures below were made from tangram A.

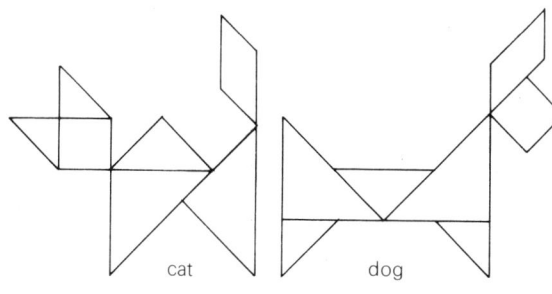

cat   dog

Tangram is probably derived from the tanka game. The tanka people lived on boats and traded with foreign ships in Chinese harbours.

TAX

An **amount** of money charged by an authority such as a Government. Examples are taxes on income, property, buying or selling goods. The money is used to pay for roads, health services, prisons and many other Government expenses.

TEENS

The **numerals** 13 to 19 inclusive.

Example: People aged 13 to 19 inclusive are said to be in their teens. They are called teenagers. So called as all of those numerals end in teen, that is they have teen as a suffix.

OLD ENGLISH (suffix) *tien*, ten.

TEMPERATURE

A **measure** of heat. Temperature is measured by means of a **thermometer**.

(See CELSIUS for a **scale** of **measurement**.)

TEMPLATE

An aid for drawing **shapes**. The required shape may be removed from a sheet of plastic or perspex as shown below.

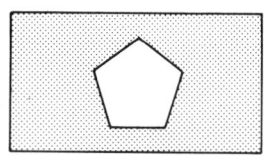

Alternatively a shape may be drawn round:

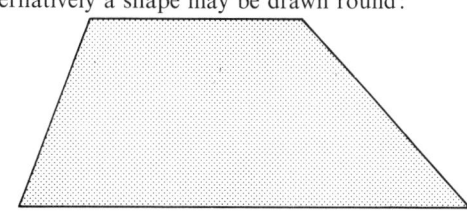

TENTH

1 The ordinal **position** after ninth.
Examples: The tenth of June. Tenth in a race.

2 One of ten **equal** parts. Written as $\frac{1}{10}$.
Example: A millimetre is one tenth of a centimetre.

TERM(S)

1 A **symbol** or **set** of symbols that are regarded as combined together.
Example: $3x^2 + 5x - 7$ consists of three terms $3x^2$, $5x$ and 7 (or $3x^2$, $+5x$ and -7).

2 See LOWEST TERMS.

3 A word that has been given a special meaning.
Example: A **square** is the term used for a **rhombus** that has a **right angle**.

TESSELLATION

A tiling **pattern**. Originally it referred only to a tiling of **equal squares** but it is now extended to any **shape** (or shapes) that repeat themselves in a regular **order** and leave no gaps.

Two types can be made from **regular polygons**:

a A regular tessellation is made with only one kind of regular polygon. The only possible sorts are made with **equilateral triangles**, squares or regular **hexagons**.

b Semi-regular tessellations use any **combination** of regular polygons, for example: a hexagon, 2 squares and an equilateral triangle.

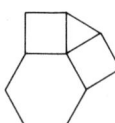

LATIN *tessalla*, small four sided tiles used to make patterns.

TETRAHEDRON

A **solid** with four **faces**.

The four faces of a **regular** tetrahedron are **congruent equilateral triangles**.

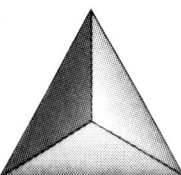

A regular tetrahedron

GREEK *tetra*, four; *hedra*, seat (face).

TETROMINOES

Shapes made by joining four **squares** so that if two meet then they have a complete **edge** in common. If one shape can be made to fit on to another by turning then the two are regarded as being the same.

Thus ⬚⬚⬚ and ⬚⬚⬚ are the same. There are five different tetrominoes.

Tetrominoes are particular cases of **polyominoes**.

GREEK *tetra*, four (ominoes is from the last part of dominoes).

THALES, 640–550 BC

Thales was a Greek businessman who became interested in **geometry** when travelling in Egypt. He laid the foundation to later Greek studies of geometry and was known as 'The Father of Geometry'. By **shadow reckoning** Thales calculated the **height** of the Great Pyramid. His method made use of **similar triangles**.

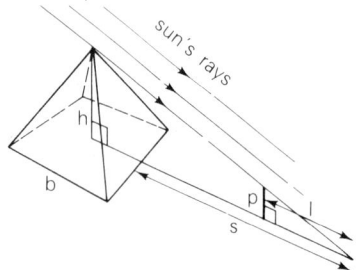

A pole, height p, had a shadow **length** ℓ.

The pyramid's **square base** had **sides** of length b. The pyramid's height was h and the length of its shadow s.

By similar triangles $\dfrac{h}{\frac{1}{2}b+s} = \dfrac{p}{\ell}$. $h = \dfrac{p(\frac{1}{2}b+s)}{\ell}$.

p, b, s and ℓ were measured and h then calculated.

THEODOLITE

An **instrument** used in **surveying** for measuring **angles**.

THEOREM

A **statement** or proposition to be proved. Now generally used for the proposition and its **proof**.

THEORY

The abstract ideas and principles concerning a body of knowledge.

Examples: The theory of **equations**. **Number** theory. **Set** theory.

THERMOMETER

An **instrument** for measuring **temperature**.

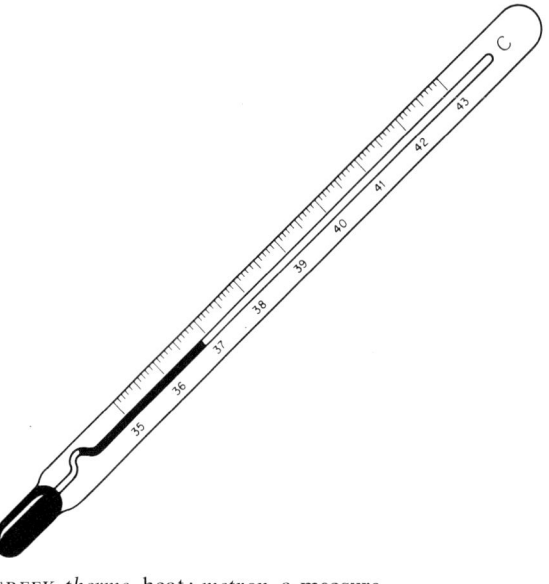

GREEK *therme*, heat; *metron*, a **measure**.

THIRD

1 The **position** in an **order** that is after **second** and before fourth.

2 A third is one of three **equal** parts. It is written as $\frac{1}{3}$.

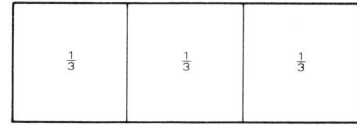

THOUSAND

Ten hundreds. Written as 1000. M in Roman **numerals**.

THREE-DIMENSIONAL

Requiring three **measurements** to **represent** its **points**. These could be **length**, **breadth** and **height**. A **solid** is three-dimensional.

TILING

See TESSELLATION.

TIME

A period of existence. Measured duration. That which elapses between one event and the next. (Sometimes referred to as the fourth **dimension** (See THREE-DIMENSIONAL.)
(See TIMES.)

TIMES

How often an **addition** is to be performed.
4 times 5 is $5+5+5+5$, that is 20.
'Times' is often used incorrectly for **multiplication**. 3 times 4 is $4+4+4$, but 3 multiplied by 4 is $3+3+3+3$. The **sign** $\times$ strictly speaking stands for 'multiplied by' and not 'times'.
3 times 4 is best written as 3(4), also read as three fours. (See TIME.)

TIME ZONES

For every $15°$ of **longitude** westwards the **time**, as measured by the sun, would be one **hour** earlier. This is because the sun moves through a complete **circle** ($360°$) in 24 hours, $(\frac{360}{24} = 15)$ i.e. $15°$ in 1 hour.
The **differences** in time could cause confusion particularly in large countries and to avoid this time zones were adopted in 1883. Each zone is contained between two **lines** running **north** to south but with adjustments to suit the **boundaries** of countries. See picture below.
Also see DATE LINE.

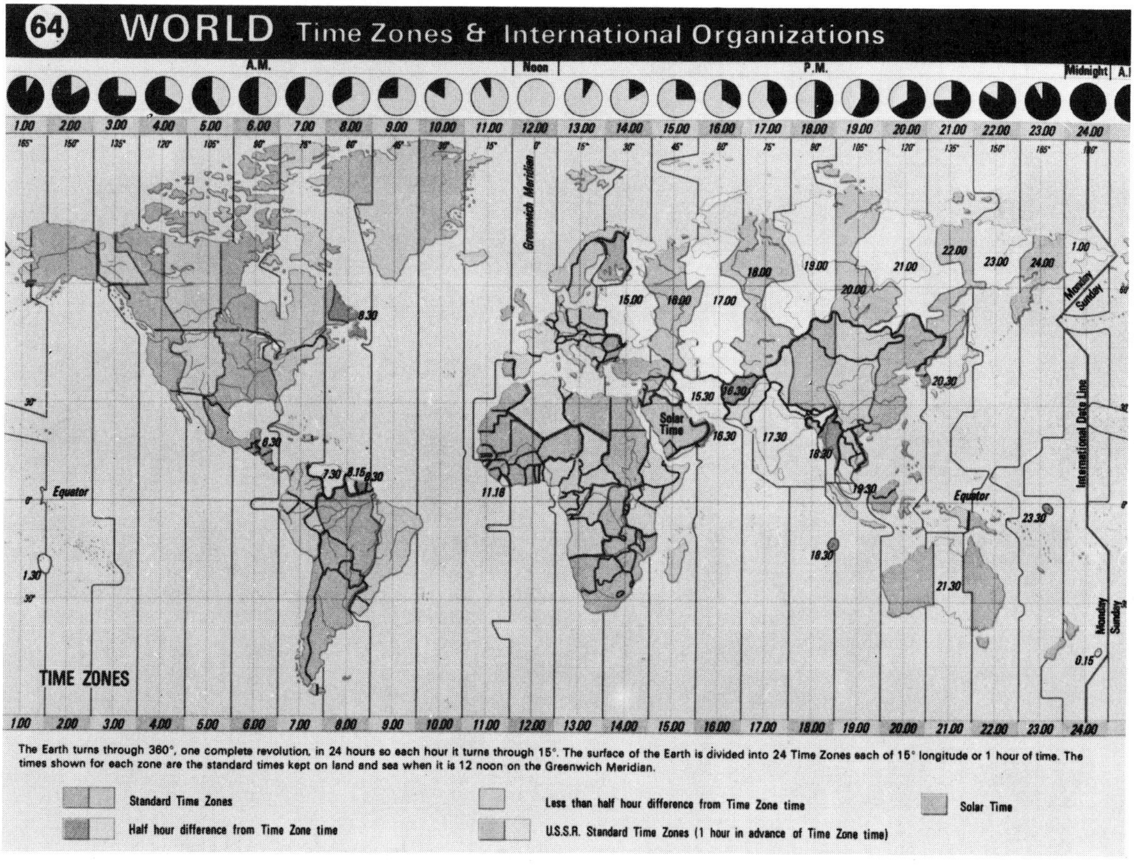

![WORLD Time Zones & International Organizations map showing world time zones with clock faces indicating hours from 1.00 A.M. to Midnight, and a world map divided into 24 time zones]

The Earth turns through 360°, one complete revolution, in 24 hours so each hour it turns through 15°. The surface of the Earth is divided into 24 Time Zones each of 15° longitude or 1 hour of time. The times shown for each zone are the standard times kept on land and sea when it is 12 noon on the Greenwich Meridian.

Standard Time Zones
Half hour difference from Time Zone time
Less than half hour difference from Time Zone time
U.S.S.R. Standard Time Zones (1 hour in advance of Time Zone time)
Solar Time

TON

An imperial **measure** of **weight**.
1 ton = 20 **hundredweights** (cwt) = 2240 **pounds** (lb).

TONNE

A metric **measurement** of **mass** or **weight**.
1 tonne = 1000 **kilograms** (kg).

TOPOLOGY

The study of the **properties** of **figures** that remain un-changed when they are stretched or bent but not torn. It is sometimes called rubber-sheet **geometry**.
Examples:
The **triangle** ABC (i) can be distorted, without tearing, to give (ii) and (iii).

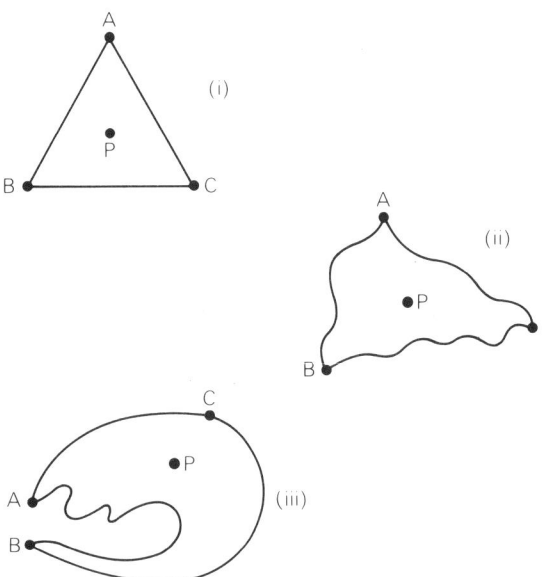

P always remains inside the figure. Whether or not a **point** is inside or outside is but one of many aspects of topology. Topology is concerned with relative **positions** but not with **measurements**.

TORUS

Also called an anchor-ring.
A **solid** made (generated) by a **closed curve** rotating around an **axis** in its own **plane**. The axis must not **intersect** the **surface**. In particular the **term** applies to a doughnut or quoit **shape** produced by rotating a **circle** about such an axis.

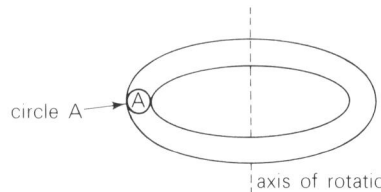

TOTAL

1 The **aggregate**. All; every one.
Example: The total population was asked to approve the new law.

2 **Sum.** Found by adding the given **values** together.
Example: The total of 18, 12, 15 and 16 is 61.

TOUCH

To be a **tangent**.

TRADITIONAL MATHEMATICS

A **term** loosely used for the **mathematics** that is not classified as **modern mathematics**.

TRAIN OF RODS

A **set** of coloured **rods** (**Cuisenaire**, **Colour factor**) placed end to end. The **diagrams** below are drawn to **scale**.

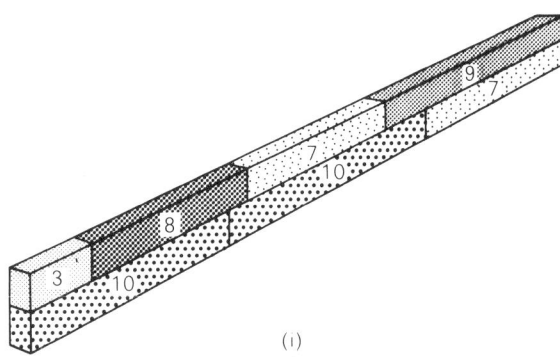

In (i) the rods are used to show **addition**. The **diagram** shows that $3+8+7+9 = 27$.

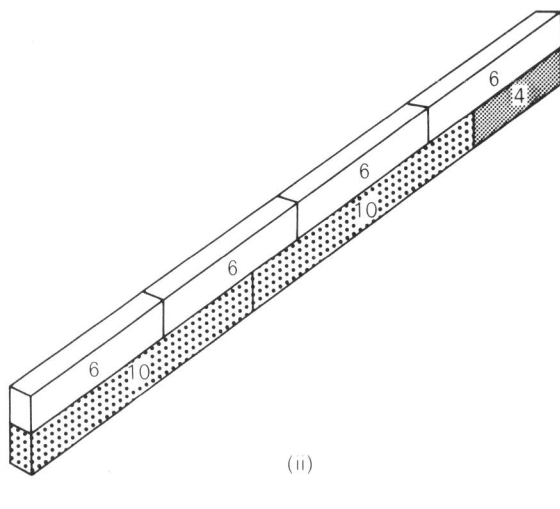

In (ii) the rods show **multiplication**. $6 \times 4 = 24$.

TRAJECTORY

The path of a **projectile** such as a stone or bullet.
Examples: The trajectory of a planet is called an **orbit**.

the trajectory of a cricket ball

LATIN *trans*, across; *jacere*, to throw.

TRANSFORMATION

1 A **mapping**.

2 In **geometry** a **one-to-one correspondence** between **points** of a **figure** in a given **position** and those of the corresponding figure in another position. **Translations, rotations, reflections** and **enlargements (dilatations)** are geometric transformations.

3 Applied to **formulae** a transformation involves expressing formulae in a different but equivalent form.
Example: $5a + 6b = c$ Express **a** in **terms** of **b** and **c**.

$$5a = c - 6b, \quad a = \frac{c - 6b}{5}.$$

This is also known as 'changing the subject of the formula'; in the example **a** was made the subject.
LATIN *trans*, across; *forma*, form.

TRANSITIVE RELATION

A **relation** such that: If (*a*) is related to (*b*) in the same way as (*b*) is related to (*c*) then (*a*) is also related to (*c*). Denoting the relation by **arrow lines**.

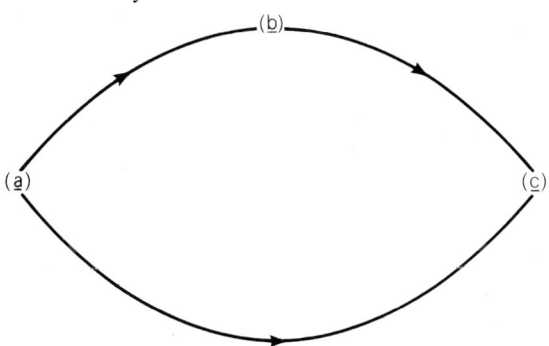

Example: Consider the **numbers** 8, 13 and 17 and the relation 'is **less than**' represented by $\longrightarrow$.
$8 \longrightarrow 13 \longrightarrow 17$. It follows $8 \longrightarrow 17$ and the relation is therefore transitive.
LATIN *transitus*, go.

TRANSLATION

A **transformation** in which every **point** of a body moves the same distance in the same **direction**.
A movement without **rotation**.

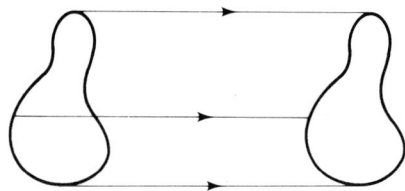

LATIN *trans*, across; *latum*, to carry.

TRANSVERSAL

A **line** that cuts across two or more other lines.
Example:
AB is a transversal.

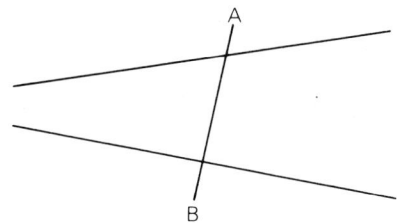

(When the other two lines are **parallel** the **alternate angles** are **equal**. Also the **corresponding angles** are then equal.)
LATIN *trans*, across.

TRAPEZIUM

A **quadrilateral** with one **pair** of **sides parallel** and the other pair *not* parallel. The plural of trapezium is trapezia.

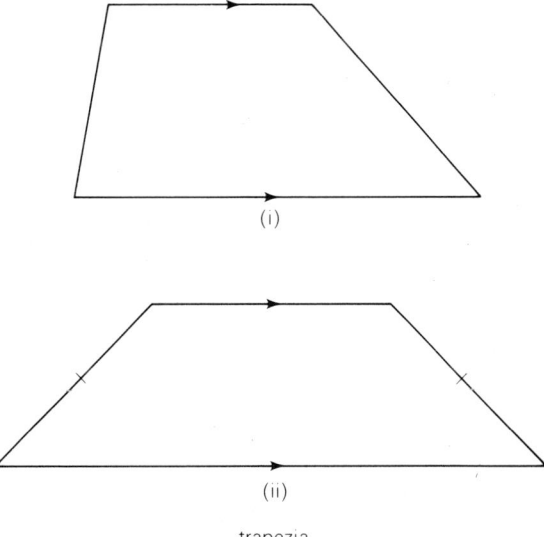

(i)

(ii)

trapezia

If the two sides that are *not* parallel are **equal**, as in (ii) the trapezium is **isosceles**.

TRAPEZOID
A **quadrilateral** with no two **sides parallel**.

TREBLE
To **multiply** by three.
Example: Treble 8 and the **result** is 24.

TRIANGLE
A **closed figure** formed by three **lines** intersecting in **pairs** or a closed figure with three **angles** and not more than three.

Examples:

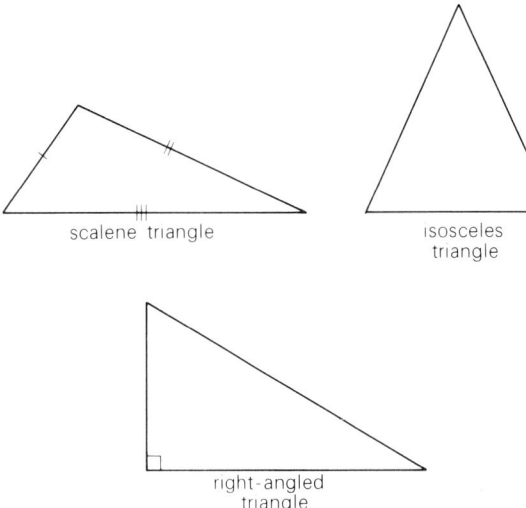

scalene triangle

isosceles triangle

right-angled triangle

The **sum** of the **interior angles** of a triangle is 180°.
LATIN *tri*, three; *angulus*, an **angle**.

TRIANGULAR NUMBERS
Numbers that can be represented by **dots** in the form of a **triangle**.
Examples:

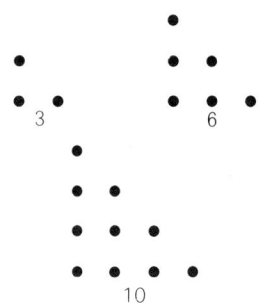

3

6

10

It is usual to include 1 as a triangular number. The first seven triangular numbers are then 1, 3, 6, 10, 15, 21, 28, ... (Note that the **difference** between one **term** and the next increases by one each **time**.)
LATIN *tri*, three.

TRIANGULATION
A method of **surveying** based on **triangles**. Starting with one triangle, say ABC, a point D is chosen and from this triangle ADC is constructed. Similarly other triangles can be added.

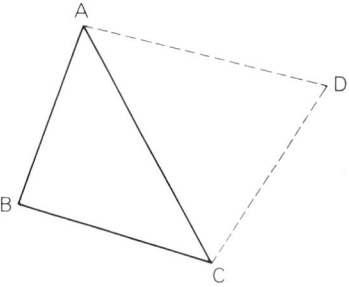

LATIN *tri*, three; **angulus**, an **angle**.

TRIGONOMETRY
The study of the **relations** of **sides** and **angles** of **triangles** and other relations that arise from these. In more advanced work this is extended to **quadrilaterals** and other **figures**.
GREEK *trigonon*, a triangle; *metron*, a **measure**.

TRIPLE
A **set** of three.
Examples: The triple (3, 5, 9) are the **coordinates** of a **point** in **three-dimensional space.**
The triple (3, 4, 12) are related in the following ways,
$3 \times 4 = 12$, $12 \div 4 = 3$, $12 \div 3 = 4$.
LATIN *triplus*, triple.

TRISECT
To **divide** into three **equal** parts.
LATIN *tria*, three; *sectum*, to cut.

TROMINO
Three **squares** joined so that any two that are joined have a complete **edge** in common. If turning is allowed there are only two trominoes.

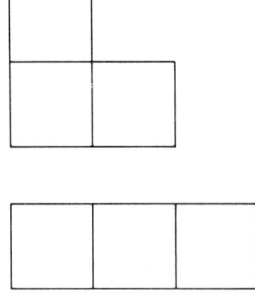

Trominoes are one particular example of **polyominoes**. Omino is from the last part of **domino** and the tr indicates 'three'.

TROY SYSTEM

A system of **weights** used for measuring precious metals or gems. The name is derived from the town of Troyes in France which was famous for jewels and other valuables. 1 **pound** troy was 5760 grains. The pound troy is no longer a legal **measure**. 1 pound troy = 12 **ounces**.

TRUE

A **sentence** or **statement** that is true when a given **value** replaces the **variable**.
Example: $2x + 3 = 9$ is a true sentence (or statement) for $x = 3$. For other values it is a **false sentence**.

TRUNDLE WHEEL

Also called a **click-wheel**. An **instrument** for measuring distances.

If the **circumference** of the wheel is 1 **metre** then for every complete turn a distance of 1 metre has been covered. The user pushes the wheel along the path to be measured. A click is heard at every complete **revolution** so that if these are counted the distance in metres is known.
OLD ENGLISH
trendel, a **circle** or anything **round**; *hweol*, wheel.

TRUTH SET

The **set** of **values** for which a **sentence** or **statement** is **true**.
Example: 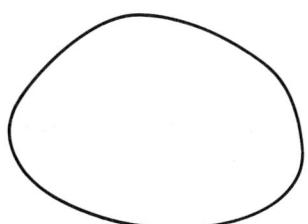 is a day of the week.
The truth set is {Sunday, Monday, Tuesday, Wednesday, Thursday, Friday, Saturday}.
Any **element** or **member** of this set, when written in the box 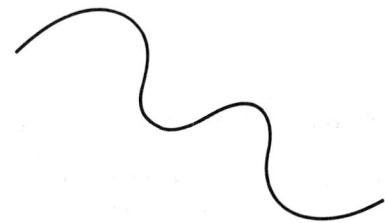 will make the sentence true.

TWO DIMENSIONS

Having **length** and **breadth**. A **surface** has two dimensions.
Examples:

The enclosed **region** is two dimensional.

A **line** has **one dimension**. In practice every line we draw has some thickness and is therefore two dimensional. (See THREE-DIMENSIONAL.)

U

U
A **symbol** for **universal set**. Now more usually written as $\mathscr{E}$ to avoid confusion with $\cup$ which indicates **union**.

UNEQUAL
Not **equal**. The **symbol** for unequal is $\neq$.
Example: $5\frac{1}{2} \times 3 \neq 15$.

UNEVEN
Not even.
Example: A road or other **surface** with bumps in it is uneven.
The **term** is not used to indicate **odd numbers** although this might seem to be logical since uneven means 'not even'.

UNIFORM
Not changing. Remaining **constant**.
Example: A uniform **speed** of 60 **kilometres** per **hour**. This means the same steady speed is maintained at all **times**.

UNION
The **combination** of two or more **sets** so as to include all their **elements** and no others. The **symbol** for union is $\cup$.
Examples:
1

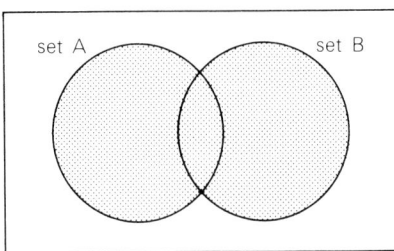

A $\cup$ B contain all elements in the shaded **region**.

2 {Boys} $\cup$ {Girls} = {Children}.
The union of a set of boys with a set of girls gives a set of children.

UNIQUE
Only one of them.
Example: The **sum** of any two **numbers** is unique. In **base** ten $18 + 5$ **equals** 23 and no other number except 23.
LATIN *unicus*, only one.

UNIT
1 A name for 'one'.
Example: 25 is 2 tens and 5 units.

2 A standard used in measuring.
Examples: The metric unit of **length** is a **metre**. A **litre** is a convenient unit for measuring **capacity**.

UNIT FRACTION
A **fraction** with 1 as **numerator** and whose **denominator** is an **integer** other than 0.
Examples: $\frac{1}{4}$, $\frac{1}{7}$, $\frac{1}{38}$.

UNITARY METHOD
A method of solving **problems** in **proportion**.
Example: 9 **kilograms** of potatoes cost 54 pence. How much will 6 kilograms cost?

9 kg cost 54p.

1 kg cost $\dfrac{54p}{9}$. (The use of 1 **unit** gives rise to the name.)

6 kg cost $\dfrac{54 \times 6p}{9} = 36p$.

UNITE
To combine together. When two **sets** are united the **term** 'union' is used.

UNITY
A name for 'one'.

UNITY ELEMENT
Another name for **identity element**.

UNIVERSAL SET

The **set** containing all the **elements** that are being considered.

Examples:

1 How many children go home to dinner? A teacher putting this question to her class would regard her class as the universal set. She does not mean how many children in the school, town or the whole country.

2 If asked to write down three **odd positive numbers** you would find no difficulty as you could select from the universal set of 1, 3, 5, 7, 9, 11, 13, ... If told the universal set was **whole numbers less than** 2 then the task is not possible.

UNKNOWN

A **symbol** used to stand for a **quantity** that is not known. In an **equation** such as $3x + 5 = 17$, x is the unknown. By solving the equation we find the **value** of x.

$3x + 5 - 5 = 17 - 5.$ $3x = 12.$ $x = 4.$

UNTRUE

Not true. False.

UP

In a **direction** from a lower place to a higher one. At or near the top as in 'He is up the ladder'.

V

V

The Roman **numeral** for five. Roman merchants **signalled amounts** to each other. The **angle** between the thumb and hand gave rise to V.

VALUE

1 The **number** resulting when an expression is simplified.

Example: The **value** of $\frac{(3+8)}{2} + 6$ is $11\frac{1}{2}$.

Find the value of $3x+2$ when $x = 6$.
$(3 \times 6) + 2 = 20$.

2 The **amount** of money an article is worth.
(See PLACE VALUE.)

VARIABLE

A **symbol** such as x which stands for an **unknown quantity**, usually a **number**.
Examples:

$3y - 1 = 8$. y is the variable.
$2 \times \square + 3 = 7$. $\square$ is the variable.

A variable acts as a **place holder**. It 'holds the place' for a number. A variable is also called an unknown.

VARIATION

A **relation** between two **sets** of **variables** according to certain rules. In direct variation the **ratio** of two variables is **constant**. (See DIRECT PROPORTION.) In **inverse** variation the **product** of two variables is constant.
(See INVERSE PROPORTION.)

VECTOR

A **quantity** that has both **magnitude** and **direction**.
Example: The **speed** and **direction** of a ship.

The direction of the vector **line** shows the direction in which the ship is travelling and its **length**, on a given **scale**, **represents** the speed.

VELOCITY

The distance travelled, in **unit time**, in a specified **direction**.
Example: 10 **kilometres** per **second** to the north-west.
Velocity differs from **speed** in that velocity has a stated direction but speed does not.

VENN DIAGRAM

A **diagram** of overlapping **shapes** used to **represent sets** and **relations** between sets.
Example: For **natural numbers less than** 20 the Venn diagram shows that the two given sets intersect with 6, 12 and 18 belonging to both sets.

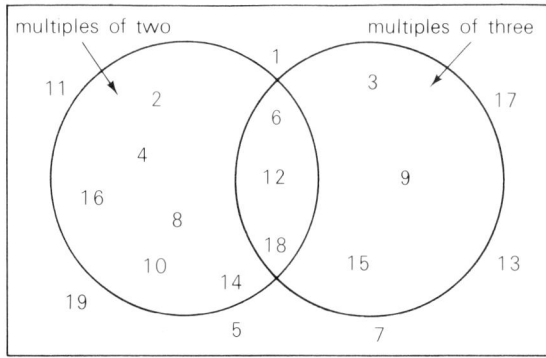

Venn diagrams are named after John Venn (1834–1923) an English mathematician who also wrote about **logic**. The diagrams are sometimes called Euler-Venn diagrams as **Euler,** a Swiss mathematician used them before Venn.

VERTEX
A **point** where two **lines** or **edges** meet. Plural vertices. *Examples:* The vertex of an **angle**.

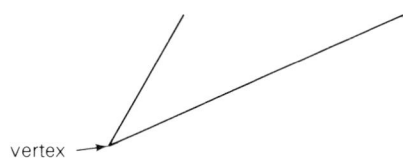
vertex

The vertices of a **triangle**.

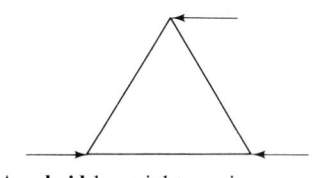

A **cuboid** has eight vertices.

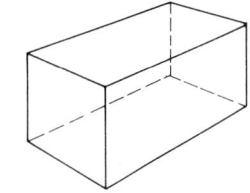

VERTICAL
A **line** that is **perpendicular** to a **horizontal** line or **plane**.

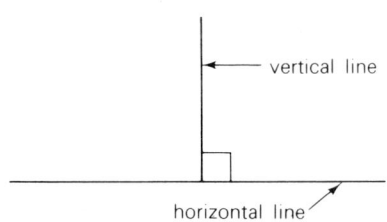
vertical line
horizontal line

In graphical work the *y*-**axis** is called the vertical **axis**. The *x*-**axis** is called the horizontal axis.

VERTICALLY OPPOSITE ANGLES
The **angles** formed where two **lines** intersect.

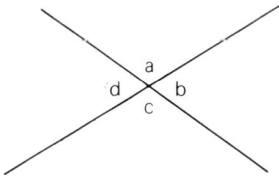
d a b c

a and c are vertically opposite angles. d and b are vertically opposite angles. Vertically opposite angles are **equal**.

VOLT
A **measure** of electric **force**. 1 volt is required to drive 1 **ampere** of current through a resistance of 1 ohm. Named after Alessandro Volta (1745–1827), an Italian scientist.

VOLUME
The **amount** of **space** occupied by a **solid**. This may be called the external volume. It is measured in cubic **units** such as cubic metres (m^3) or cubic centimetres (cm^3). The amount of space inside a solid (such as a box) is called the internal volume (or cubic **capacity**).
[Note: The **term** solid is still used even if there is space inside.]

VULGAR FRACTION
See **common** or **simple fraction**.
LATIN *vulgaris*, the common (ordinary) people; **fractum,** to break.

W

WATT
A **unit** of electrical **power**. The **symbol** is W.
1 **kilowatt** (1000 watts) is approximately $1\frac{1}{3}$ horse-power.
(1 watt = 1 joule per **second**).
Named after James Watt (1736–1819).

WEIGHT
The **force** with which a body is pulled towards the Earth's **centre**.
Weight is not the same as **mass**. If a body moves into **space** its mass is not changed but its weight becomes less as it moves farther from the Earth because the force of attraction becomes less. When the forces of attraction from the Earth, moon and other bodies balance the body will be weightless, but its mass is unchanged.

WEEK
1 Seven days.

2 A working week generally applies to the five days Monday, Tuesday, Wednesday, Thursday and Friday.

WELL-DEFINED SET
A **set** so described that it is clear as to whether any **element** is, or is not, in the set.
Examples: Vowels in the English alphabet (well-defined). A fair person (not well-defined). A **whole number less than** ten (well-defined).

WHOLE NUMBERS
The positive **integers** 0, 1, 2, 3, 4, 5, ... **Zero** together with all the **natural numbers**.

WIDTH
See BREADTH.

WRONG
Not right. Incorrect. Answers are not always simply **right** or wrong. A **measurement** may be given as, say, 28.3 centimetres. With more accurate **instruments** it may be given as 28.32 cm and even more accurately as 28.324 cm.

No instrument can **measure** perfectly and hence no measurement is completely right. They are all given to a certain degree of accuracy.

XYZ

X, x

1 The Roman **numeral** for ten.

2 The **sign** for **multiplication**. Came from the Greek letter chi χ.

X-AXIS

The **horizontal axis** in **Cartesian coordinates**.

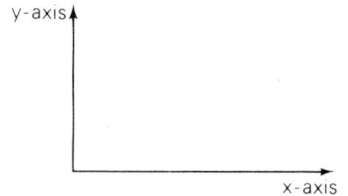

YARD

A **unit** of **length** in the **imperial system**.

1 yard = 3 feet = 36 **inches**.

1 yard is approximately 0.91 **metres**.

In Northern Europe a yard was the length of the girdle the Anglo-Saxons wore. In southern countries it was the length of a **double cubit** (2×18 inches). At the start of the 12th **century** Henry I fixed the yard as the distance between his nose and the thumb of his outstretched arm. The yard before 1439 was 39.66 inches, very close to a metre.

y-AXIS

The **vertical axis** in **Cartesian coordinates**.

YEAR

The year you are familiar with is also called the astronomical, equinoctial, natural or solar year. The period of **time** it takes the Earth to make a revolution around the sun. It is 365 days, 5 **hours**, 48 **minutes** and 45.7 **seconds**. A sidereal year is determined by the **positions** of the stars and is complicated to calculate so we do not use it for normal **calculations**.

ZERO

1 The **cardinal number** of an **empty set**. There are no **elements**.
Nothing, **nought** (or **naught**).

2 A **place holder**, as in 500.

3 The **identity element** in **addition**.
FRENCH *zero*, zero (from Arabic *sifr*, **cipher**, **nothing**).

TABLE OF SYMBOLS

Symbol		*Example*
+	Plus, add, positive	$3+8=11$
−	Minus, subtract, negative	$7-2=5$
×	Multiplied by, times	$5 \times 4 = 20$
÷	Divided by	$12 \div 2 = 6$
/ −	Divided by. Denoting a fraction	$3/4 = \frac{3}{4}$ (Three-quarters)
$\sqrt{}$	Square root	$\sqrt{9} = 3$
$\sqrt[3]{}$	Cube root	$\sqrt[3]{27} = 3$
∠ ∧	Angle	$\angle A = 30°$, $\hat{B} = 45°$
△	Triangle	$\triangle ABC$ is right-angled
——	Line segment	$\overset{A}{\vdash}\!\!\!—\!\!\!\overset{B}{\dashv}$ AB is a line segment
↗	Vector	↗ represents the wind strength and direction.
⊥	Perpendicular	AB is perpendicular to CD
‖ ⟹	Parallel	$PQ \parallel RS$
°	Degree	Boiling point for water is 100°C.
		The angle was 48°
=	Equals	$4+4=8$
≠	Is not equal to	$9-2 \neq 6$
<	Is less than	$7 < 8$
>	Is greater than	$6 > 3$
⩽	Is less than or equal to	$x \leqslant 3$. x can be 3 or any number less than 3
⩾	Is greater than or equal to	$x \geqslant 4$. x can be 4 or any number greater than 4
≮	Is not less than	$7 \not< 4$

$\not>$	Is not greater than	$5 \not> 7$
$\approx \ \simeq \ \fallingdotseq$	Is approximately equal to	$\sqrt{2} \approx 1.414$
$\cong \ \equiv$	Is congruent to	Triangle ABC $\cong$ triangle PQR
$\cap$	Intersection, cap	The shaded part shows the intersection of sets A and B $^A \ \text{⦾} \ ^B$
$\cup$	Union, cup	The shaded part shows the union of sets A and B. $^A \ \text{⦿} \ ^B$
$\{\ \}$	Braces enclosing a set	The set of vowels is $\{a, e, i, o, u\}$
$(\ )$	Brackets showing which part	$13 - [5 + 2(4 + 1) - 3] = 13 - [5 + 8 + 2 - 3]$
$[\ ]$	to evaluate first	$= 13 - [12] = 1$
$\therefore$	Therefore	$\hat{A} = \hat{B}, \ \hat{A} + \hat{B} = 90°$ $\therefore \ \hat{A} = \hat{B} = 45°$
$\because$	Because	$\hat{A} = 30°, \ \hat{A} = \hat{B} \ \because \hat{B}$ is also known to be $30°$
$:$	such that, is to	$\{x : x > 5\}$ The set of values for x such that x is greater than five.
$::$	as	$2 : 4 :: 8 : 16.$ 2 is to 4 as 8 is to 16, $(\frac{2}{4} = \frac{8}{16})$
$\subset$	Is included in Is a subset of	$\{2, 4, 8, 12, 20\} \subset \{\text{even numbers}\}$
$\not\subset$	Is not included in Is not a subset of	$\{1, 3, 5, 11, 21\} \not\subset \{\text{even numbers}\}$
$\supset$	Contains, includes the subset	$\{\text{even numbers}\} \supset \{4, 8, 12\}$
$\mathcal{E}, \text{E}, \mathcal{U}$	Universal set	Find all the even numbers less than 100. The universal set being $\{0, 1, 2, 3, \ldots 100\}$
$\%$	per cent	5 per cent of 40 is 2
$'$	minute	$1°$ contains $60'$. An angle of $1'$ is therefore $\frac{1}{60}$ of $1°$
$''$	Second	An angle of $1'$ is equivalent to $60''$. For angles 1 second is $\frac{1}{60}$ of a minute
$\propto$	Varies as	The perimetrer (p) of a square varies at the length (l) of its side, p $\propto l$.
∞	Infinity	The sum of all the counting numbers is infinity, $1 + 2 + 3 + \cdots = \infty$
$\Rightarrow$	Implies	$x = 3 \Rightarrow x^2 = 9$
$\Leftrightarrow$	Implies and is implied by	$2x = 4 \Leftrightarrow x = 2$
$\varnothing, \{\ \}$	Empty set	$\{\text{Triangles with four sides}\} = \varnothing$
W	Set of whole numbers	$\{0, 1, 2, 3, 4, 5, \ldots\}$
Z	Set of integers	$\{\ldots -2, -1, 0, 1, 2, \ldots\}$

N	Set of natural numbers	$\{1, 2, 3, 4, 5, \ldots\}$
Q	Set of rational numbers	All integers are rational numbers, $Z \subset Q$
R	Set of real numbers	Rational numbers are a subset of the real numbers, $Q \subset R$

ENGLISH LETTERS USED IN MATHEMATICS

a, b, c	Sides or lengths. Constants in algebra.
A	Area
C, c	Circumference of a circle
d	Length of diameter
h	Height
p	perimeter
r	Length of radius
V, v	Volume

GREEK LETTERS USED IN MATHEMATICS

Many of these have several uses, especially in Mathematics at a higher level than we are concerned with. Only a few of the main explanations are therefore given.

α	Alpha	
β	Beta	
γ	Gamma	
Δ	Delta	
ϵ	Epsilon	Is a member, or element, of a set
$\notin$	stands for 'is not a member'	
θ	Theta	
π	pi	Approximately 3.1416 or $3\frac{1}{7}$

The value $\dfrac{\text{circumference}}{\text{diameter}}$ for a circle.

$\Sigma \, \sigma$	Sigma
ϕ	Phi
μ	Mu

USEFUL TABLES

Length, linear measure

10 millimetres (mm)	= 1 centimetre (cm)
10 centimetres	= 1 decimetre (dm)
10 decimetres	= 1 metre (m)
10 metres	= 1 decametre (dam)
10 decametres	= 1 hectometre (hm)
10 hectometres	= 1 kilometre (km)
1000 mm = 1 m	1000 m = 1 km

Area, square measure

100 square millimetres (mm^2)	= 1 square centimetre (cm^2)
100 square centimetres	= 1 square decimetre (dm^2)
100 square decimetres	= 1 square metre (m^2)
100 square metres	= 1 square decametre (dam^2)
100 square decametres	= 1 square hectometre (hm^2)
100 square hectometres	= 1 square kilometre

1 are (a) = 100 square metres
100 ares = 1 hectare = 10 000 m^2
1 square kilometre = 100 hectares = 1 000 000 m^2

Volume, capacity. Cubic measure

1000 cubic millimetres (mm^3)	= 1 cubic centimetre (cm^3)
1000 cubic centimetres	= 1 cubic decimetre (dm^3)
1000 cubic decimetres	= 1 cubic metre (m^3)

1 litre (ℓ) = 1000 cm^3 = 1 dm^3

10 millilitres (mℓ)	= 1 centilitre (cℓ)
10 centilitres	= 1 decilitre (dℓ)
10 decilitres	= 1 litre (ℓ)
10 litres	= 1 decalitre (daℓ)
10 decalitres	= 1 hectolitre (hℓ)
10 hectolitres	= 1 kilolitre (kℓ)

Weight, mass

10 milligrams (mg)	= 1 centigram (cg)
10 centigrams	= 1 decigram (dg)
10 decigrams	= 1 gram (g)
10 grams	= 1 decagram (dag)
10 decagrams	= 1 hectogram (hg)
10 hectograms	= 1 kilogram (kg)
1000 kilograms	= 1 tonne (t)

Angles

60 seconds (") = 1 minute (')
60 minutes = 1 degree (°)
90 degrees = 1 right angle (∟)
180 degrees = 1 straight angle
360 degrees = 1 complete turn

Time

1000 milliseconds = 1 second (s)
60 seconds = 1 minute
60 minutes = 1 hour (h)
24 hours = 1 day
7 days = 1 week
365 days = 1 year (366 days = 1 leap year)
12 calendar months = 1 year
10 years = 1 decade
100 years = 1 century

(There is no standard abbreviation
for minute but min is acceptable.
m should *not* be used as it is
the abbreviation for metre.)

SOME USEFUL FORMULAE

PLANE SHAPES

Circle

(radius r, diameter d)

Circle $\quad C = \pi d = 2\pi r$

Area $\quad A = \pi r^2 = \dfrac{\pi d^2}{4}$

Parallelogram

Area $\quad A = bh$

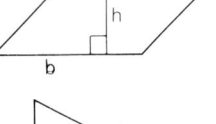

Pythagoras' Theorem

$$a^2 = b^2 + c^2$$

Rectangle

Area $\quad A = ab$

Perimeter $\quad p = 2(a+b)$

Square

Area $\quad A = s^2$

Perimeter $\quad p = 4s$

Trapezium

Area $\quad A = h\left(\dfrac{a+b}{2}\right) = \tfrac{1}{2}h(a+b)$

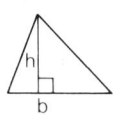

Triangle

Area $\quad A = \tfrac{1}{2}bh$

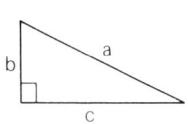

168

SOLIDS

Cube

Volume $\qquad V = s^3$

Surface Area $\quad A = 6s^2$

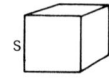

Cuboid

Volume $\qquad V = abc$

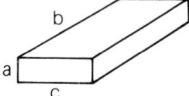

Pyramid

Volume $\qquad V = \frac{1}{3}$ Area of base $\times$ height

$\qquad\qquad V = \frac{1}{3}Ah$

Right circular cone

Volume $\qquad V = \frac{1}{3}Bh$ (B is area of base)

$\qquad\qquad = \frac{1}{3}\pi r^2 h$

Right circular cylinder

Volume $\qquad V = \pi r^2 h$

Area of curved surface $A_1 = 2\pi rh = \pi dh$

Total surface area $\qquad A_2 = 2\pi r(r+h)$

Sphere

Surface area $\quad S = 4\pi r^2$

Volume $\qquad V = \frac{4}{3}r^3$

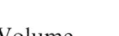

Simple interest

I = Interest, P = Principal, T = Time,

R = Rate, A = Amount

$I = \dfrac{PTR}{100} \qquad A = P + I$

169